W9-DCU-663

WITHDRAWN

# ELECTRICAL DRAFTING

**ERROL G. SCHRIEVER**

Electrical Design Supervisor
S.I.P. Engineering, Inc.
Houston, Texas

and

Vocational and Technical Instructor
San Jacinto College
Pasadena, Texas

**PRENTICE-HALL, INC.**

Englewood Cliffs, New Jersey 07632

*Library of Congress Cataloging in Publication Data*

Schriever, Errol G.
Electrical drafting.

Includes index.
1. Electric drafting. I. Title.
TK431.S37 1984 621.319'24'0221 83-4521
ISBN 0-13-247288-0

Editorial/production supervision
and interior design: Kathryn Gollin Marshak
Cover design: Photo Plus Art—Celine Brandes
Manufacturing buyer: Anthony Caruso



Printed in the United States of America

10 9 8 7 6 5 4 3 2 1

ISBN 0-13-247288-0

Prentice-Hall International, Inc., *London*
Prentice-Hall of Australia Pty. Limited, *Sydney*
Editora Prentice-Hall do Brasil, Ltda., *Rio de Janeiro*
Prentice-Hall Canada Inc., *Toronto*
Prentice-Hall of India Private Limited, *New Delhi*
Prentice-Hall of Japan, Inc., *Tokyo*
Prentice-Hall of Southeast Asia Pte. Ltd., *Singapore*
Whitehall Books Limited, *Wellington, New Zealand*

*To my wife and family,*
*whose enthusiasm and support enabled me*
*to complete this project,*
*I dedicate this text.*

# CONTENTS

# PREFACE

As the demand for electrical energy continues to grow throughout the world, so increases the demand for knowledgeable individuals to depict these installations in a concise, professional manner. This is the field of electrical drafting.

In the early stages of electrical drafting, a knowledge of electrical theory, although helpful, is not absolutely necessary. Of prime importance is a basic understnding of the methods and techniques used in the industry and the symbology used to represent the various aspects necessary to the drawing and to the finished product. This is the purpose of this text.

As you advance in your career in electrical drafting, you will learn basic theory by osmosis. Interest in your work and the desire to do your best will aid this learning process. This is not to say that you should not avail yourself of every opportunity to do additional study in electrical theory. This you should do in addition to the work experience you are involved in.

This book is not a design manual, although some of the criteria covered approach basic design. It is an effort to acquaint the individual, who is already familiar with the basic drafting fundamentals of line work and lettering with the practices and procedures used in the layout and construction of an electrical drawing. In addition, this text is geared primarily to the role of the electrical drafter in an industrial complex. For purposes of familiarity, a section on commercial wiring and methods is included.

There are numerous self-check quizzes throughout the chapters. These are spotted so that you can check your progress as you study. Cover the right side of the quiz, the answer portion, and fill in the blanks; then check your answers. Section references are given to check your remarks.

An appendix is included, which provides tables and other data that will aid in solving problems and in basic design.

Every effort has been made to prepare this text as a basic tool for those going into electrical drafting or those in the early stages of their careers. It is the sincere wish of the author that this text will serve the purpose for which it was intended.

The author sincerely appreciates the untiring efforts of Kimberly Knight, whose skill with the typewriter and ever-present smile made this project a pleasure instead of a task. The Square-D company graciously furnished the information on starter and control terminology. Their kindness is also appreciated.

*Errol G. Schriever*

# BASIC PROCEDURES

## 1.1 INTRODUCTION

The drafting of electrical installations requires the same basic fundamentals of good lettering and line work as do all other disciplines. It is not absolutely necessary for the electrical drafter to know the intricate workings of the drawing being done, and the theory behind it, but it is necessary to see that the drawing is neat, legible, and accurate. Additional knowledge can be gained from every drawing assigned. The drafter should study the drawing, attempt to follow the circuitry, and formulate reasons for the designer's choice of product or routing. In other words, at all times draw with your head instead of your pencil. Know generally what you are drawing; if you do not, ask questions.

## 1.2 DRAWING BOARD COVERING FOR ELECTRICAL DRAWINGS

Since many electrical drawings are in the form of diagrams, schedules, or tables, it is helpful to cover the drawing board surface with a smooth-surface graph paper made especially for this purpose. When the blank drawing sheet is placed on the board over the graph paper covering, the grid lines are visible and therefore very helpful as a guide. A common-sized graph paper is made in 1-in. grids divided into $^1/_8$-in. squares. Figure 1.1 shows this size.

The grid paper in Figure 1.1 is often light green in color, and the surface finish is glossy so that tape or "dots" are easily removed without disturbing the surface. When the grid paper is put on the board, care must be taken to be sure the lines of the grid line up exactly with the edge of the parallel bar or drafting rule. The paper can then be secured with masking tape and replaced as required.

## 1.3 DRAWING TO SCALE

Most electrical layout drawings are drawn to scale. Accurate work with the scale is important for the designer and drafter as a means of checking clearances, space requirements, and the like, to the contractor because he usually has to scale the drawings to

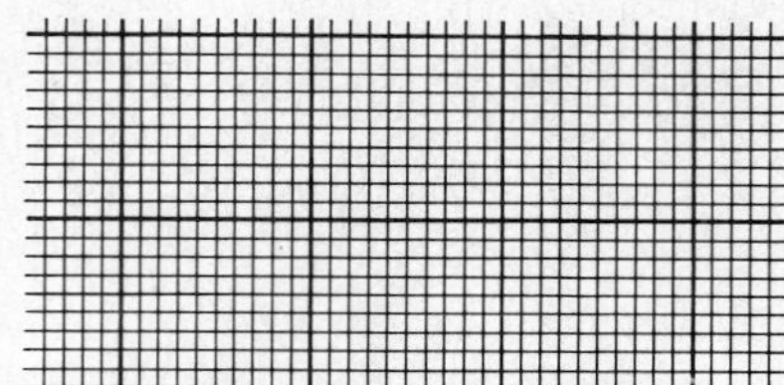

**FIGURE 1.1.** Graph Paper Table Covering

determine material requirements, and also to the installing electrician in the field as a guide to locating equipment and so on. A note is often included stating *not* to scale on the drawings. This is done because scaled dimensions cannot be guaranteed as accurate, and therefore a worker must be careful of interferences if he or she uses a dimension scaled from a print.

Usually, electrical layouts are not drawn to scales smaller than $^1/_8$" = 1′-0″. If smaller scales are used, the information to be shown probably covers a large area, and a larger scale would require dividing the work up on several drawings. To avoid this, the smaller scale can be used, and where details need to be developed, smaller *part* or *partial plans* can be shown at larger scales. Many electrical drawings are made at the scale of $^1/_4$" = 1′-0″. This usually permits the necessary amount of room for detail and is a fairly easy scale to use. As the scale becomes larger, the drafting is easier, so the largest scale that can be used practically is to be desired.

## 1.4 CHOICE AND USE OF TEMPLATES

There are several templates made expressly for use in drafting electrical drawings. Most of these templates include standard symbols used in wiring diagrams. The technique in the successful use of templates is to be careful to follow the symbol cutout closely and to keep the pencil lead sharpened to the right size for drawing a neat and clean-cut line. If the pencil is too dull, it may not fit into the cutout opening at all; if it is too sharp, it may slide around in the cutout and cut the paper. Also, care must be taken to be sure the template is properly in line on the drawing to assure that the symbol will be correctly placed.

If a template is used, the symbols can be constructed in a few simple steps. For example, an incoming line symbol can be made from a triangle in the template; transformer windings are a series of half-circles, safety switches and motor starters are made by using squares, and disconnect devices are made with the triangle cutouts. Some of these simple symbols are shown in Figure 1.2, using the template shown. This is a standard do-all template. One type of electrical template, of which there are many, is shown in Figure 1.3.

## 1.5 DRAWING CHARACTERISTICS

Any drawing going to the field for any reason should be of good quality, concise, accurate, and complete. Electrical drawings are no exception. Line work and lettering should be sharp, good, and representative. Notes and statements should be complete, understandable, and as simple as possible. The drawing should be accurate, and finally the drawing should be free of smudges and dirt. A smudge on a tracing can completely cover up an important dimension or word on a print. Cleanliness is a sign of the pride that drafters take in their work. The drawing subject should be the first thing seen. If it is a conduit plan, the conduit runs stand out; if it is a lighting plan, the first things seen are fixtures.

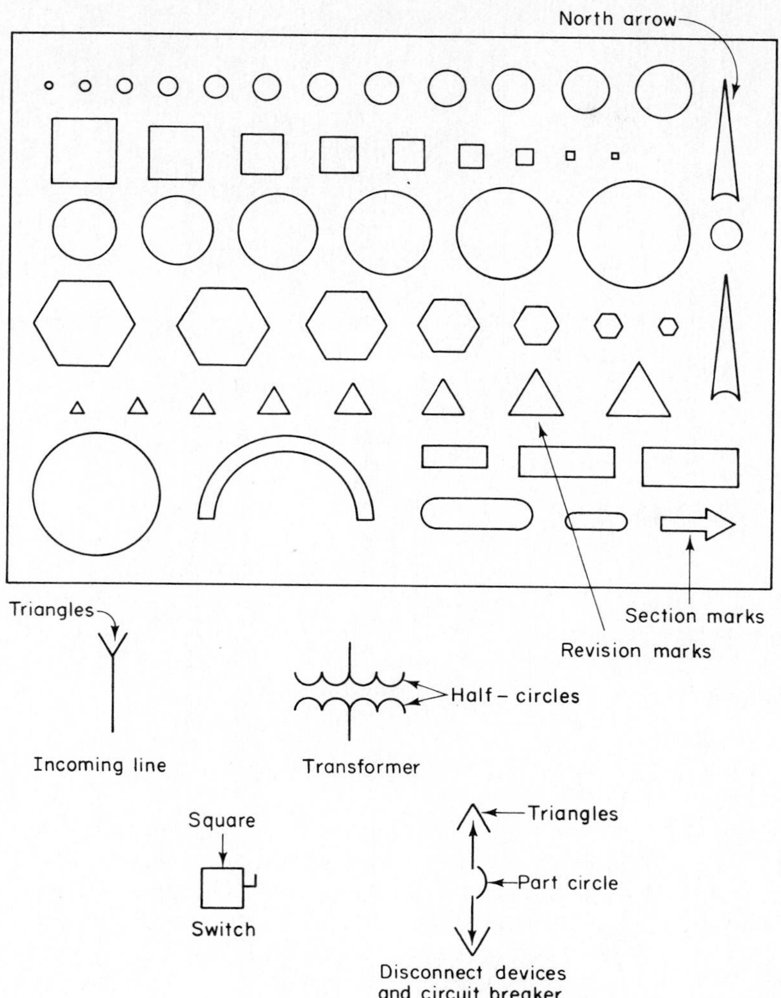


**FIGURE 1.2.** Standard Do-All Template

## 1.6 PLANT COORDINATE SYSTEM

### 1.6.1 Definition

Just as the earth has latitudes and longitudes to enable locations to be expressed mathematically, almost all industrial installations utilize a grid system, which is called a *plant coordinate system*. To define a particular spot on the earth, two coordinates are given, such as latitude $X$ and longitude $Y$. This establishes two straight lines 90° apart, referred to as *cross hairs*. The two coordinate lines meet at a point that is unlike any other point on the earth, and so it becomes a precise location.

### 1.6.2 Bench Mark

Plant coordinate systems use north–south and east–west lines as coordinates. A *bench mark,* or base point, is established, and coordinate lines are assigned to this point. The bench mark must be something that will remain stable; it is usually a concrete marker firmly planted in the ground. As an example, the bench mark may be assigned the coordinates of north 0.0′ and east 0.0′. An item assigned coordinates of east 10.0′ and north 150.0′ would have its cross hairs located 10′ east of the bench mark and 150′ north of the bench mark.

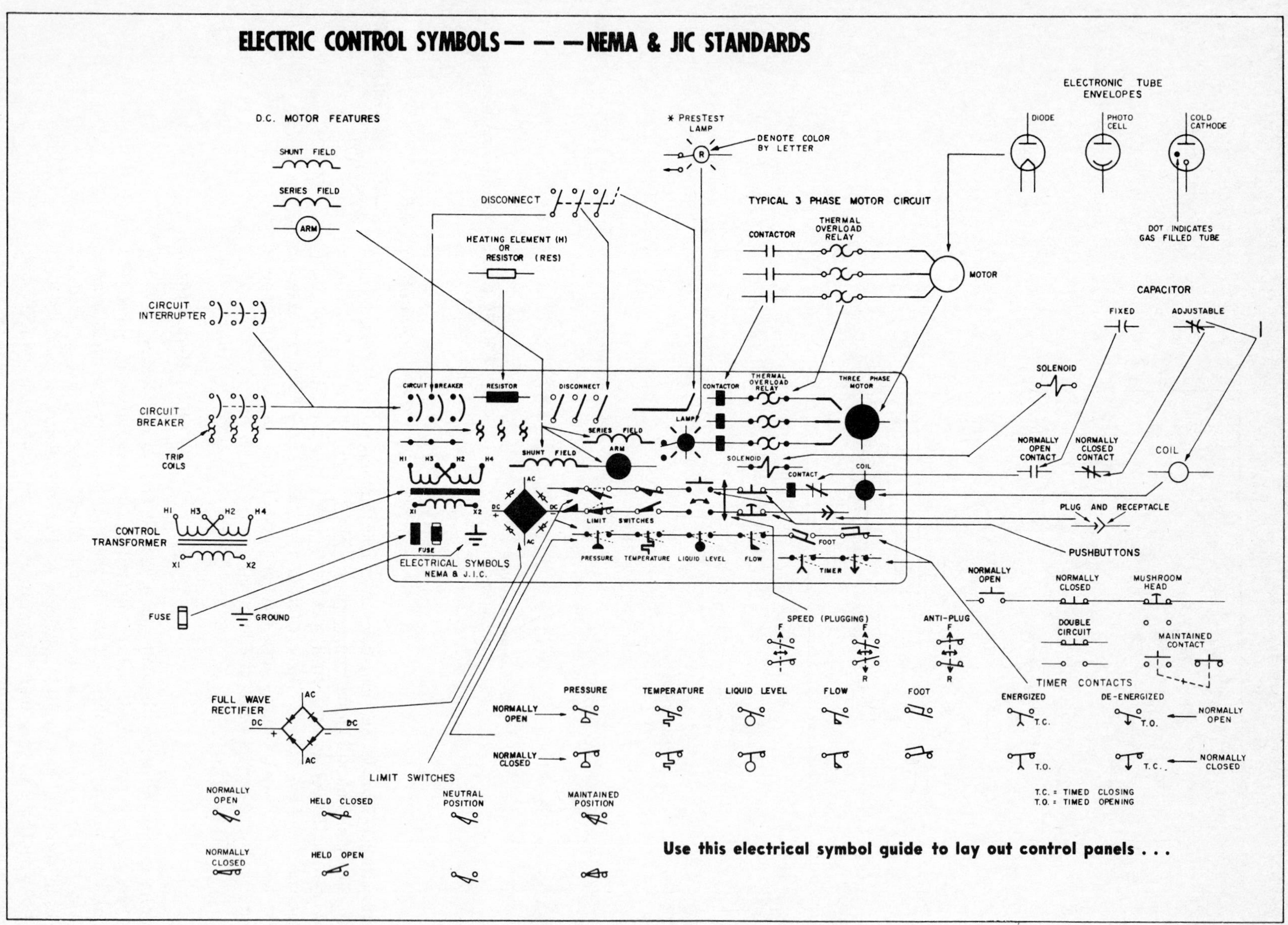


**FIGURE 1.3.** Electrical Template

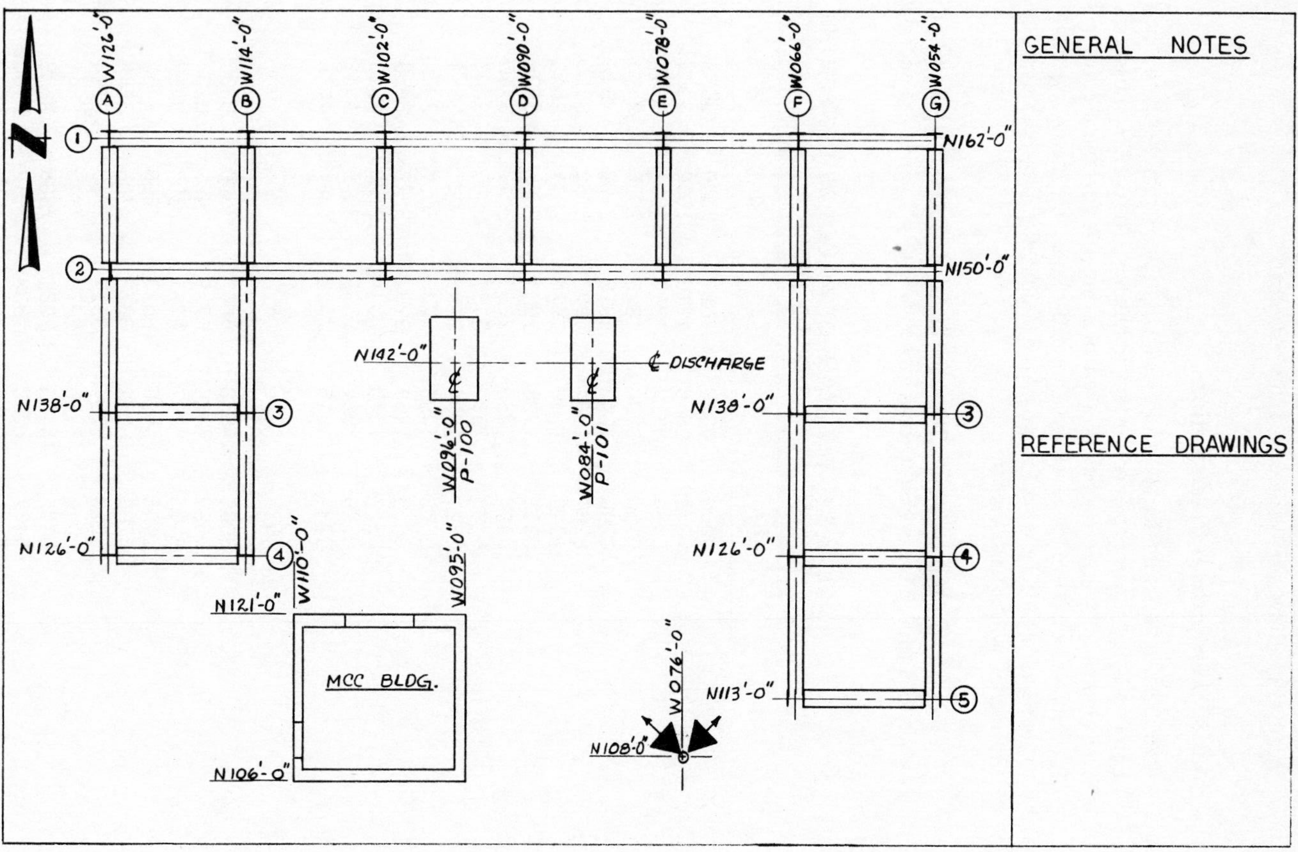


**FIGURE 1.4.** Use of Plant Coordinates

The bench mark is also assigned a plant elevation. This elevation might be elevation 100.00′. In some plants the actual elevation above sea level is used at the bench mark. Usually, the mythical elevation of 100 is given to bench marks. Why don't plants call the bench mark elevation 0.0′? This would cause items located below grade to have *minus* elevations, such as −6.50, and this would be confusing and could cause errors. Minus elevations are avoided in industrial plants.

### 1.6.3 Methods of Presentation

Overall, plant coordinates are usually shown in the decimal system, such as N. 1438.63′. Plant drawings are dimensioned in feet and inches. The drafters must convert these decimal coordinates to feet and inches on their drawings. A few plant coordinate systems are in feet and inches; however, since most are in decimals, the methods used to convert decimals to feet and inches must be known. An example of a plot plan using coordinates in feet and inches is illustrated in Figure 1.4.

***Decimal Conversions.*** The coordinate N. 150.00′ is expressed in feet and decimals of a foot. This is easily converted to N. 150-0″, since the decimal shown was .0′.

**Table 1.1 DECIMALS OF A FOOT BY SIXTEENTHS OF AN INCH**

| | *0″* | *1″* | *2″* | *3″* | *4″* | *5″* |
|---|---|---|---|---|---|---|
| 0″ | .0000 | .0833 | .1667 | .2500 | .3333 | .4167 |
| 1/16″ | .0052 | .0855 | .1719 | .2552 | .3385 | .4219 |
| 1/8″ | .0104 | .0937 | .1771 | .2604 | .3437 | .4271 |
| 3/16″ | .0156 | .0990 | .1823 | .2656 | .3490 | .4323 |
| 1/4″ | .0208 | .1042 | .1875 | .2708 | .3542 | .4375 |
| 5/16″ | .0260 | .1094 | .1927 | .2760 | .3594 | .4427 |
| 3/8″ | .0312 | .1146 | .1979 | .2812 | .3646 | .4479 |
| 7/16″ | .0365 | .1198 | .2031 | .2865 | .3698 | .4531 |
| 1/2″ | .0417 | .1250 | .2083 | .2917 | .3750 | .4583 |
| 9/16″ | .0469 | .1302 | .2135 | .2969 | .3802 | .4635 |
| 5/8″ | .0521 | .1354 | .2188 | .3021 | .3854 | .4688 |
| 11/16″ | .0573 | .1406 | .2240 | .3073 | .3906 | .4740 |
| 3/4″ | .0625 | .1458 | .2292 | .3125 | .3958 | .4792 |
| 13/16″ | .0677 | .1510 | .2344 | .3177 | .4010 | .4844 |
| 7/8″ | .0729 | .1562 | .2396 | .3229 | .4062 | .4896 |
| 15/16″ | .0781 | .1615 | .2448 | .3281 | .4115 | .4948 |
| | *6″* | *7″* | *8″* | *9″* | *10″* | *11″* |
| 0″ | .5000 | .5833 | .6667 | .7500 | .8333 | .9167 |
| 1/16″ | .5052 | .5885 | .6719 | .7552 | .8385 | .9219 |
| 1/8″ | .5104 | .5937 | .6771 | .7604 | .8437 | .9271 |
| 3/16″ | .5156 | .5990 | .6823 | .7656 | .8490 | .9323 |
| 1/4″ | .5208 | .6042 | .6875 | .7708 | .8542 | .9375 |
| 5/16″ | .5260 | .6094 | .6927 | .7760 | .8594 | .9427 |
| 3/8″ | .5312 | .6146 | .6979 | .7812 | .8646 | .9479 |
| 7/16″ | .5365 | .6198 | .7031 | .7865 | .8698 | .9531 |
| 1/2″ | .5417 | .6250 | .7083 | .7917 | .8750 | .9583 |
| 9/16″ | .5469 | .6302 | .7135 | .7969 | .8802 | .9635 |
| 5/8″ | .5521 | .6354 | .7188 | .8021 | .8854 | .9688 |
| 11/16″ | .5573 | .6406 | .7240 | .8073 | .8906 | .9740 |
| 3/4″ | .5625 | .6458 | .7292 | .8125 | .8958 | .9792 |
| 13/16″ | .5677 | .6510 | .7344 | .8177 | .9010 | .9844 |
| 7/8″ | .5729 | .6562 | .7396 | .8229 | .9062 | .9896 |
| 15/16″ | .5781 | .6615 | .7448 | .8281 | .9115 | .9948 |

If this had been N. 150.50′, it is apparent that .50′ is one half of a foot or 6″, so N. 150.50′ becomes N. 150′-6″.

With a coordinate of N. 150.76′, conversion to feet and inches is not quite so apparent. To convert decimals to feet and inches, drafters refer to a conversion table. Table 1.1 supplies these conversions.

To convert N. 150.76′, it is known that there are 150′. The unknown is how many inches, plus fractions of an inch, are in the .76′. Referring to Table 1.1, .76′ is located under the 9″ column, across from $^1/_8$″ on the left side. So, .76′ is converted to $9^1/_8$″. The fully converted coordinate then becomes N. 150′-$9^1/_8$″.

### 1.6.4 Examples

Often, the drafter may know two coordinates and need to know the distance between them. *If the smaller coordinate is subtracted from the larger, the difference is the distance between them.* It must be noted that this applies only to like coordinates, such as subtracting a north coordinate from another north coordinate, as shown in the following examples.

*EXAMPLE:*

```
     N. 1824.73′
   − N. 1638.92′
        185.81′
   =    185′-9 3/4″
```

*EXAMPLE:*

```
     E.  278.3724′
   − E.  123.7269′
         154.6455′
   =     154′-7 3/4″
```

The reverse of this method is used to establish a coordinate when one coordinate is known and the distance between the known and unknown coordinate is known. If the known coordinate is N. 183.5622′ and the known dimension is 13′-$2^3/_8$″ north of this coordinate, then:

1. Convert 13′-$2^3/_8$″ to feet and decimals of a foot. Using Table 1.1, 13′-$2^3/_8$″ equals 13.1979′.
2. Since 13′-$2^3/_8$″ is north of the known coordinate, the answer must be larger than the N. 183.5622′. In other words, the coordinate is farther north from the bench mark. So 13.1979′ must be added to the known coordinate.

*EXAMPLE:*

```
     N. 183.5622′
   +     13.1979′
     N. 196.7601′     is the unknown coordinate.
```

Had the known dimension, 13′-$2^3/_8$″, been south of the known coordinate, N. 183.5622′, it would be closer to the bench mark and the dimension, 13.1979′, would have to be subtracted from the known coordinate.

*EXAMPLE:*

```
     N. 183.5622′
   −     13.1979′
     N. 170.3643′     is the unknown coordinate south of the known
                      one.
```

Normally, a plant will not have both north and south coordinates. Occasionally they will, so the student must remember that each *coordinate is a dimension of a certain distance from the bench mark.* If N. 18′-0″ is 18′-0″ north of the bench mark, then S. 18′-0″ is 18′-0″ south of the same bench mark. The dimension between N. 18′-0″ and S. 18′-0″ is 36′-0″. The rule is, *to find the distance between north and south coordinates, add the two numbers.*

*EXAMPLE:*

| | | |
|---|---|---|
| | N. 138.6623′ | |
| + | S. 36.5277′ | |
| | 175.1900′ | = 175′-2¼″ is the dimension between coordinates |

The same rule applies when both east and west coordinates are given in the same plant.

North and south coordinates have been used in this example, but they could just as easily have been east and west coordinates. The same rules apply. A dimension 10′-6″ east of E. 18.5′ would make the unknown coordinate E. 29.0′.

*EXAMPLE:*

| | | |
|---|---|---|
| | E. 18.5′ | |
| + | 10.5′ | (10′-6″ converted to decimals) |
| | E. 29.0 | |

Since the 10′-6″ dimension was east of the known coordinate, the unknown coordinate was farther east of the bench mark, so 10′-6″ is added to get a larger east coordinate than E. 18.5′. Had the dimension 10′-6″ been west of E. 18.5′, it would be closer to the bench mark and would have to be subtracted from E. 18.5′.

*EXAMPLE:*

| | | |
|---|---|---|
| | E. 18.5′ | known coordinate |
| − | 10.5′ | known dimension |
| | E. 8.0′ | unknown coordinate |

### 1.6.5 Plant Elevations

A plant coordinate system is based on everything being on a flat plane. A plant's plot is considered to be flat for coordinate purposes, although it actually may be hilly. Vertical dimensions are referred to as *elevations*.

An elevation point above sea level is established at the bench mark. This elevation may be 10′, 150′, 500′, or 2218′ above sea level, but to keep plant numbers consistent an arbitrary elevation is usually assigned. This is usually 100′-0″. Any elevation referred to as, say, 92′-3″ is known to be below grade. Most companies supply the actual elevation above sea level on each drawing by a note that may read, "elevation 100′-0″ equals actual elevation 1275′-3¼″.

With this system, it is easily recognized that a pipe rack with elevation 116′-0″ TOS called out means that the top of steel is 16′-0″ above grade. A platform at elevation 133′-6″ is 33′-6″ above grade.

Piping is often located by a BOP (bottom of pipe) elevation. Since all pipe fabrication works with the pipe center line, the piping drafter must convert BOP's to center-line dimensions. To do this, the pipe ODs must be memorized. If a pipe was resting on a rack that was at elevation 116′-0″ TOS, the center-line elevation would be figured by adding one-half of the pipe OD (outside diameter) to the TOS elevation of 116′-0″. If the pipe was 3″ size, 3½″ OD, 1¾″ would be added to elevation 116′-0″, making the pipe center-line elevation equal 116′-1¾″.

**SELF-CHECK QUIZ 1.1** **Cover the right side of the page and answer the questions.**

|  |  |
|---|---|
| 1.1 Most electrical drawings are ____________ ____________ ____________. | 1.1 Ans.: *Drawn to scale* Ref. 1.3 |
| 1.2 As a rule, electrical layouts are not drawn to scales smaller than ____________" = 1′-0". | 1.2 Ans.: $^1/_8$" Ref. 1.3 |
| 1.3 ____________ is a sign of pride by the drafter in his or her work. | 1.3 Ans.: *Cleanliness* Ref. 1.5 |
| 1.4 A plant coordinate system is a ____________system. | 1.4 Ans.: *Grid* Ref. 1.6.1 |
| 1.5 Plant coordinate systems use ____________–____________ and ____________–____________ lines as coordinates. | 1.5 Ans.: *north–south* *east–west* Ref. 1.6.2 |
| 1.6 ____________ elevations are avoided in industrial plants. | 1.6 Ans.: *Minus* Ref. 1.6.2 |
| 1.7 Coordinates are usually shown in the ____________system. | 1.7 Ans.: *Decimal* Ref. 1.6.3 |
| 1.8 Plant drawings are dimensioned in ____________and ____________. | 1.8 Ans.: *feet* and *inches* Ref. 1.6.3 |

## 1.7 DIMENSIONING AND LABELING

Dimensions on electrical drawings should be referenced to column center lines, building lines, center line of major equipment, finished floor, and the like. Dimensions to outside surfaces of columns, walls, and similar surfaces with variable positive position should be avoided. Dimensions up to and including 24 in. are often given in inches and larger dimensions are given in feet and inches, although this varies from company to company.

Care should be taken to avoid dimensions to small fractions of an inch. An electrician, in the field, can seldom set a device or piece of equipment to as close a tolerance as 1/16" in. Sometimes dimensions just do not add up even if they relate to one another unless small fractions appear, but these fractions should be avoided if possible. The easiest and fastest method for dimensioning is to place the dimension above the dimension line rather than to leave a break in the line for the dimension. Labels to identify lines, equipment, and the like, should be located outside dimension lines and *never* across them.

The use of long lines with arrowheads for labeling is preferred to crowding labels into congested space, and very often this will also permit organization of several labels in a group, which is an aid in reading the drawing. As an example of this method, Figure 1.5 shows several electrical conduits connecting into equipment mounted on a wall. By running long leader lines outside the congestion, the conduit runs are kept clear of any additional confusion.

## 1.8 CHECKING DRAWINGS

Before a drawing is signed and issued for construction, it must be thoroughly checked. Checking procedures vary; however, a system that seems to work quite well is to draw a yellow line through anything that checks all right, a red line indicating an error requiring correction, and a green line through anything to be removed. The checker may choose to cover the area that is wrong with red and mark over with black pencil the correction to be made. If the correction is extensive, the checker may circle it in red and make a note explaining the error and correction to be made. Sometimes an attached sketch can be used. Some checkers use a blue pencil to note reference data used in their checking.

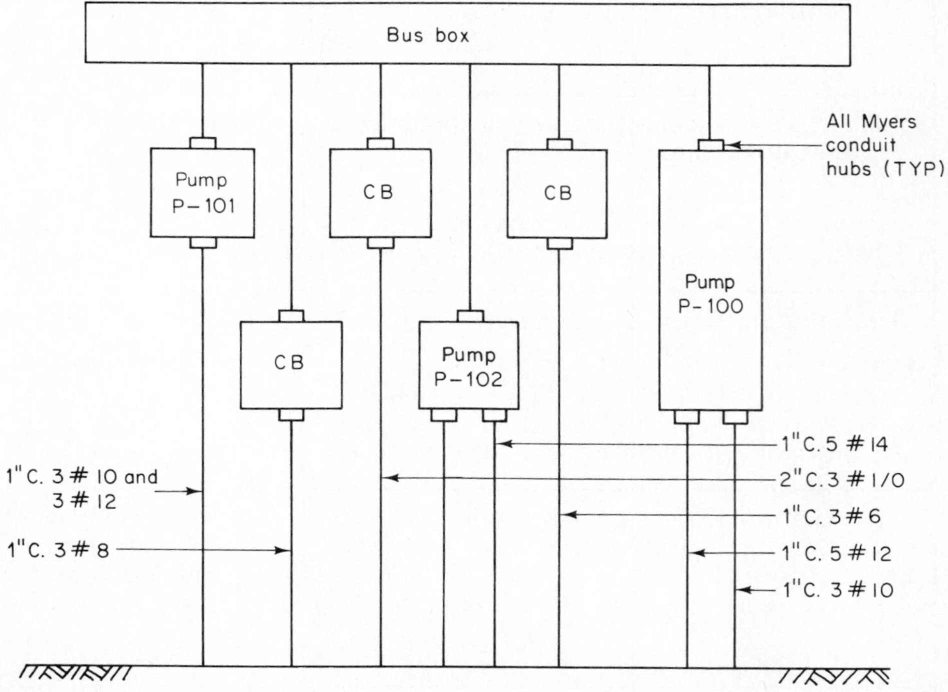


**FIGURE 1.5.** Correct Method of Dimensioning

Whatever method is used, the checking changes and instructions should be clear so that corrections can be readily made and understood.

Good checking requires a conscientious, methodical, step-by-step screening of all the information on the drawing. A costly interference in the field can be avoided by careful checking; therefore, this responsibility should not be taken lightly. A list for checking follows:

1. Uses latest reference material.
2. Understands design concepts.
3. Drawing data agree with data from other related electrical drawings.
4. Spelling.
5. Proper abbreviations (spell words out if possible).
6. Dimensions.
7. Interference of other designs (structural, mechanical, etc.)
8. Clearances sufficient for installation, removal, maintenance, and the like, of equipment.
9. Arrowheads directed to correct reference point.
10. Conduit sizes and numbers agree with schedule if schedule is used.
11. Size of wire and cables.
12. Size of conduit and other raceway.
13. Not more than three 90° bends in conduit runs. (Code does allow four or a maximum of 360° between pull points or fittings. This includes *all* bends and offsets).
14. Uniform lettering.
15. Views properly identified and referenced.
16. Scales indicated.
17. North arrow shown, if required.
18. Notes read well, stand alone, easily understood.
19. Correct title and drawing number.
20. Correct references and reference drawings.
21. Drawing is clean and legible.

*A checker should never take anything for granted.* Improvements in design should be suggested but a checker should not try to redesign an accepted design.

## 1.9 NATIONAL ELECTRIC CODE®*

The National Electric Code® is a code of standard practices and safety requirements published and distributed by the National Fire Protection Association, NFPA, in Quincy, Massachusetts. Revised and reissued every three years, it is the most widely adopted code of practices and requirements in the world. The primary purpose of the code is protection of life and property.

Originating in 1897, there have been 41 editions printed since that date. The NFPA has been distributing the code since 1951. The complete text of the National Electric Code® is found in volume 6 of the NFPA National Fire Codes. The NEC® is used by engineering personnel and electrical contractors for the basic design and installation of

*The National Electric Code® and NEC® are Registered Trademarks of the National Fire Protection Association, Inc., Quincy, Massachusetts.

electrical equipment. The code is also used by local inspection authorities responsible for safe electrical installations and by instructors teaching electrical design, drafting, and maintenance courses.

As stated, the basic and primary purpose of the NEC® is the protection of personnel and property from the many hazards present in the use of electricity. It contains provisions considered necessary by panels of experts for safety in electrical equipment and device installation.

### 1.9.1 Code Arrangement

The NEC® is divided into an introduction and nine chapters. Chapters 1 through 4 apply generally. Chapters 5 through 7, while supplementing the general rules, apply to special occupancies, special equipment, or other special conditions. Chapter 8 covers communication systems and is independent of the other chapters except where they are specifically referenced therein. Chapter 9 contains tables and examples of code applications.

### 1.9.2 Installations Covered by the NEC®

The conditions and requirements set forth in the NEC® apply to:

1. All electrical conductors and equipment installed in or on public or private buildings or structures, industrial substations and plant sites, recreational vehicles, and other premises.
2. Conductors that connect these installations to a source of electricity.
3. Other outside conductors and equipment on the premises.

### 1.9.3 Installations Not Covered by the NEC®

In spite of the wide acceptance of the NEC® provisions, a number of installations are exempt from coverage:

1. Installations in ships, railroad rolling stock, aircraft and automotive vehicles.
2. Installations underground in mines.
3. Installations by railways for generation, transformation and distribution of power for their own use.
4. Installation of communication systems or equipment used specifically by the communication utilities.
5. Installations by electric utilities for the purpose of generation, control, transformation, metering, transmission, and distribution of electric energy.

### 1.9.4 Enforcement

It was never the intent of the NFPA to accept responsibility for the enforcement of the provisions outlined in the NEC®. Local governmental bodies, through their inspection departments, exercise this responsibility of interpretation and enforcement. Such inspection groups also have the authority to modify, change, or waive various requirements and allow alternate methods of installation and construction. The prime consideration in any deviation from the code is that such alternative method(s) will not compromise the original objectives of safety set out by the NEC®. Nationally recognized testing laboratories, inspection agencies, or other organizations concerned with product evaluation publish lists of equipment or materials that have been tested and meet nationally recognized standards or that have been found suitable for use in a specified manner. The code does not contain detailed information on equipment or materials, but refers to the products as *listed, labeled,* or *approved for the purpose*. The code does not apply to the internal factory-installed wiring, or to the construction of listed equipment at the time of installation, unless damage or alterations are detected.

Members of 23 panels are responsible for changes, revisions, additions, and deletions to the National Electric Code®. These panels encompass over 400 representatives from all phases of the electrical engineering, medical, construction, administrative, and supply industries in all parts of the United States. Article 70-XIV of the code gives a list of the organizations involved on the panels, and in the front of the code book is a list of the panel members and the articles for which they are responsible. As soon as one edition is published, work starts on the next.

## 1.10 GENERAL NOTES

On most drawings some items, conditions, or requirements are more easily conveyed by words than by lines. If a certain condition is imposed on several similar items on the drawing it is quicker to write a note to cancel the whole drawing condition than to spell it out by each item, device, or the like. These notes are called *general notes* and appear in the right-hand column of the drawing sheet.

As an example look at Figure 10.12. In the right-hand column at the top of the sheet is a list of sentences headed General Notes. Note 3 says, "All underground conduit runs shall be schedule 40 heavy wall PVC and shall be encased in red concrete min. 3″ cover on all sides." It is much easier and quicker to cover the requirements with this general note than to indicate this on each conduit run in the underground duct banks.

General notes are exactly what the name implies, general information to make the drawing more understandable and complete and to cover special conditions.

## 1.11 REFERENCE DRAWINGS

Also on the right-hand side of Figure 10.12 is the heading Reference Drawings. Under this heading are listed the drawing numbers of all drawings in the package that are referred to on this drawing or that will aid in a better understanding of this drawing.

## 1.12 CHECKS AND ISSUES

During the course of a project, there are several checks and issues that a drawing or sets of drawings must be subjected to.

### 1.12.1 Squad Check

In any project, regardless of size, in which more than one group, individual, or discipline is involved, there must be an open line of communication and thought between these two or more parties for a satisfactory completion of the project and to avoid conflicts in design. This interchange of information is brought about through the squad check, whereby one discipline will send a print of a drawing or a set of drawings to all others involved in the project. They in turn check, comment on, and return it to the issuing party. These comments are documented by the sender. A drawing may be sent out for a squad check several times during the course of a project.

### 1.12.2 Electrical Check

After the design and drafting are virtually complete on a drawing, it is sent to another member of the department for checking. He or she checks all phases of the drawing, references, notes, catalog numbers, workability, design and layout, continuity, and all other aspects. The checker then initials the drawing and returns it to the sender, who studies and records any comments. See Section 1.8.

### 1.12.3 Approval Issue

Before the drawings are issued to a contractor, they are sent to the client for approval. This is known as an approval issue. The client has the final say on design, placement, layout, sequence of operation, and so on, and makes any comments or changes on this set. The drawings, with comments, are then returned to the sender, the comments are recorded, and any necessary additions, revisions, or changes are made.

### 1.12.4 Bid Issue

After the design is complete, several contractors are selected to submit bids for the construction of this facility. Complete sets of drawings are sent to these contractors, who in turn estimate material and total hours and submit their bid to the client. This issue of drawings is the bid issue. In some cases, where every phase of engineering is complete and no further changes or additions are anticipated, this issue may also be a construction issue.

### 1.12.5 Construction Issue

This is the final issue of all drawings to the contractor so that he may build the facility.

In *all* cases it is wise to keep the copies of the various squad checks, electrical checks, approval and bid issues, and the construction issue for your records.

## 1.13 CHANGES AND REVISIONS

The revision block or area located, generally, above the title block in the lower left-hand corner of the sheet is used to record issues and changes to that drawing. Looking at the sample title area in Figure 1.6, we see that there is provision for the following:

a. Revision numbers
b. Date of revision
c. Brief description of work done
d. Initials of individual doing work
e. Space for checker's initials
f. Space for approved initials

| 2 | As – built | Rm | Es | Jd | 4-16-80 |
|---|---|---|---|---|---|
| 1 | General revisions | Rm | Es | mp | 4-1-80 |
| O | Issued for construction | Rm | Es | Rf | 2-10-80 |
| A | Issued for approval | Rm | Es | www | 11-2-79 |
| No. | Revision | Dr | Ck | App | Date |
| (A) | (C) | (D) | (E) | (F) | (B) |

**FIGURE 1.6.** Revision Block

## 1.14 REVISIONS PRIOR TO CONSTRUCTION ISSUE

Revision identification prior to the issuance of the drawing for construction shall be alphabetical beginning with A. A notation in the revision block should appear for squad check issues, approval issues, and so on, with letters of the alphabet as identification.

***Construction Issue and After.*** When a drawing is issued for construction, that issue identification is 0 and revisions after that are numerical.

***Scoping.*** When a revision is made on a drawing that has been issued, a cloud is drawn on the back of the drawing completely enclosing that portion of the drawing that was changed. In this cloud is a triangle with the revision number inside.

Only the cloud for the latest revision will remain on the drawing. As a new revision is made, the old cloud is removed and the new change noted. See Figure 1.7.

***Drawing Registers.*** In addition to the drawing, there is also a column on the drawing register where this revision is recorded. This maintains a current register and a check to verify that the prints used are the latest issue.

## 1.15 ENGINEERING STANDARDS

### 1.15.1 Definition and Coverage

Standards, generally called engineering standards, are exactly what the name implies. They are standard practices, sometimes pictorial, sometimes narrative, that have been developed and perfected over a period of time. Companies may issue standards for any area in any discipline. Such standards give others insight into the methods and procedures employed to accomplish a given task or installation and open the avenue of similarity throughout the company.

There are no defined areas to be covered by company standards. They may be issued to cover the simplest task, such as a standard for replacing the plastic liner in a trash disposal container (wastebasket), to standard procedures for programming an IBM computer. The difficulty or complexity of a task is not important. The method or procedure is the concern in standard practices.

Enforcement varies from company to company. In some cases, upper management has decreed that all work performed by or for the company shall be per approved company standards, while other companies merely use them as guidelines or samples.

### 1.15.2 Advantages and Disadvantages

Engineering standards provide a very effective tool for engineering and construction provided they are studied prior to use and it is verified that the reference installation will coincide exactly with the standard used. In too many cases, additional confusion has been caused by referring to a standard and not noting exceptions or differences between the standard and the actual installation. If in doubt about the validity of a standard, it is better to draw a detail of the installation and make it part of the drawing package.

Some engineering firms issue full-sized drawings of client standards, five or six to a sheet, and then refer to these sheets in the balance of the package. This is acceptable provided the standard matches the installation.

## 1.16 DRAWING REGISTER

The drawing register can easily be considered one of the most important documents in any project. It is a listing of all the drawings required of a discipline, showing drawing number and complete title for each drawing. It is continually updated to reflect the progress made on drawings by marking the percent complete.

The drawing register will have the following:

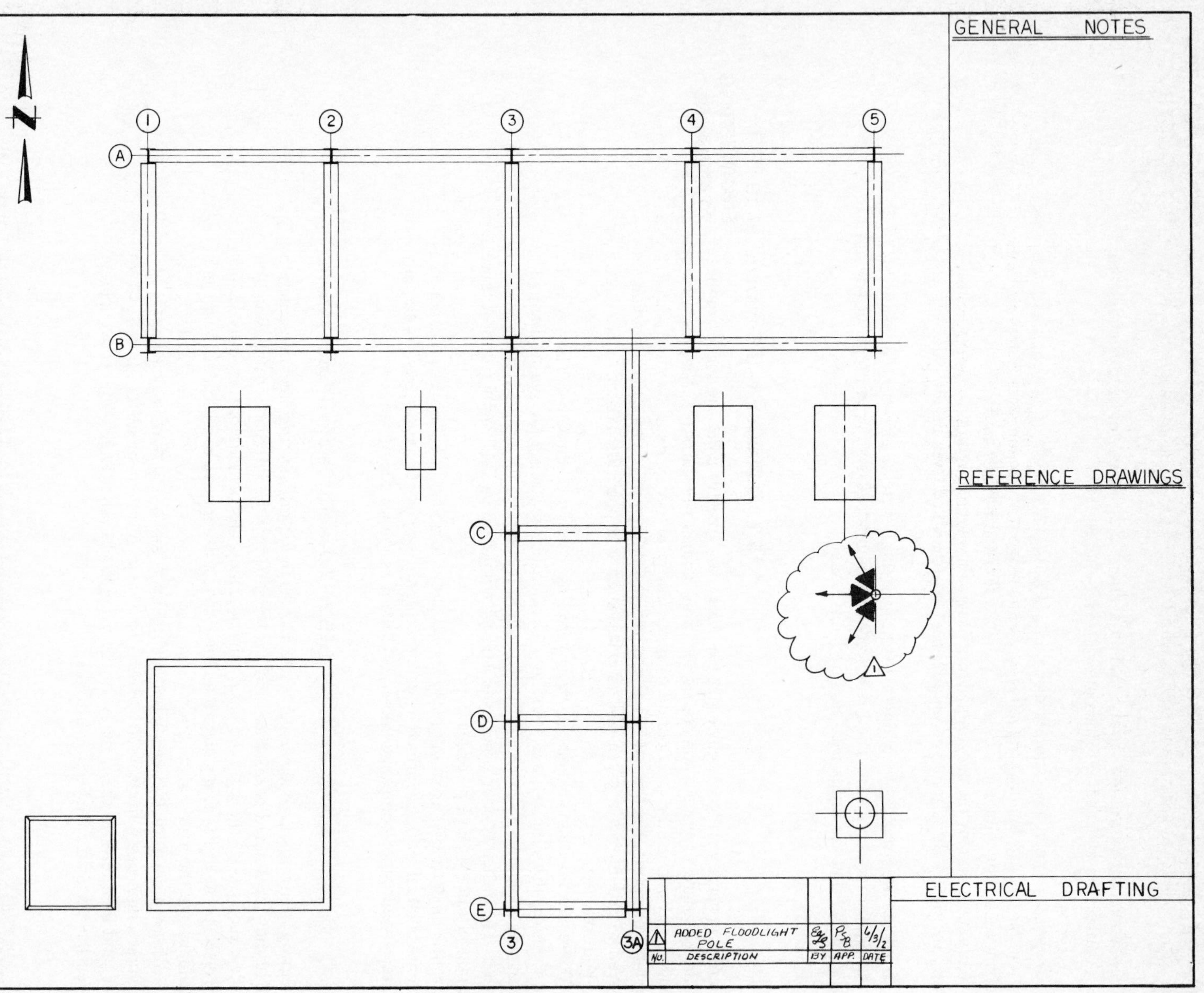


**FIGURE 1.7.** Drawing Revision

**SELF-CHECK QUIZ 1.2** **Cover the right side of the page and answer the questions.**

| | |
|---|---|
| 1.9 With two like coordinates, if the smaller coordinate is subtracted from the larger coordinate, the difference is the ________ ________ ________. | 1.9 Ans.: *Distance between them* Ref. 1.6.4 |
| 1.10 Vertical dimensions are referred to as ________. | 1.10 Ans.: *Elevations* Ref. 1.6.5 |
| 1.11 To find the distance between north and south coordinates, ________the two numbers. | 1.11 Ans.: *Add* Ref. 1.6.4 |
| 1.12 Dimensions should be referenced to ________positive position surfaces. | 1.12 Ans.: *Fixed* Ref. 1.7 |
| 1.13 The national electric code is distributed by the ________ ________ ________ ________. | 1.13 Ans.: *National Fire Protection Association* Ref. 1.9 |
| 1.14 The national electric code has ________chapters. | 1.14 Ans.: *9* Ref. 1.9.3 |
| 1.15 An interdiscipline check of a drawing is called a ________check. | 1.15 Ans.: *Squad* Ref. 1.12 |
| 1.16 Standards promote ________in plant installations. | 1.16 Ans.: *Similarity* Ref. 1.14.1 |

1. *Drawing numbers (typed).*
2. *Drawing titles.* Where drawing titles have not been assigned, a general description is used for each drawing of each category. As drawing titles are established, these general descriptions are replaced with correct titles. (Correct titles are typed, whereas general descriptions are handwritten.)
   a. Examples of general descriptions:
      Plot Plan A, or One-line Diagram A
      Plot Plan B, or One-line Diagram B
3. *Scheduled issue date.* This date must comply with the project schedule. If this date has to be changed due to design changes, project schedule revisions, and the like, the change must be approved. If a change occurs, the first date is marked through and the new date is written in the second row.
4. *Approval issue and return dates.* This records the date the drawings are given to the drafting coordinator by the design supervisor for transmittal to the client and the date of the return of the consolidated comments from the client. The approval issue should indicate revisions with a letter (A, B, C, etc.).
5. *Construction issue date and revision number.* This records the date and the revision number for the *first* issue of the drawing for the indicated project. The data record when the drawing was given to the drafting coordinator for issue. The revision number is 0 for the new drawings, and 1, 2, 3 or the next consecutive number for revisions to existing drawings (including other project drawings).
6. *Revision number and date.* This records the revisions to the drawings for the project beyond the initial construction issue. The first revision number (the actual number might be 37) and the date go in the first row, the second in the second row, the fifth will replace the first revision, and so on, depending on the number of lines available for revision (see Figure 1.6).
7. *General comments*
   a. The drawing number, drawing title, and information at the top and bottom of the page except for the issue dates must be typed. All other information will be handwritten with a soft lead pencil, generally an F or an H grade.
   b. It is preferred that only one drawing type be listed on each page, for examples, process flowsheets, structural steel, piping plans and sections, plot plans, or electrical drawings. A page or set of pages is used for each.
   c. See Figure 1.8 for a typical drawing register and status.

## 1.17 DRAWING STATUS

The primary purpose and intent of this form is to provide, for the design supervisor, a technique or tool for calculating the status of the drawings for his or her discipline. The following information is obtained:

1. Total hours estimated to be used on project.
2. Total hours, estimated or calculated, that have been spent toward completion of the drawings.
3. In the latter stages of the project, the total estimated hours to complete.

With this information, management can report to the client a relatively accurate status of the engineering and design phase and can evaluate personnel requirements necessary to complete the project within the alloted time. In addition, the overall status of the

## DRAWING REGISTER

CLIENT__________

PROJECT TITLE__________

CLIENT JOB NO.__________

MONTHLY REPORT DATES

| DRAWINGS ISSUED THIS TRANSMITTAL | DRAWING NO. | DRAWING TITLE | SCH. ISSUE | APPVL. ISSUE / RETURN | CONST. ISSUE | REVISION | |
|---|---|---|---|---|---|---|---|
| | | | | | | NO. | DATE |

JOB NO.__________ DWG. TYPE__________ PAGE__________

## DRAWING STATUS

CLIENT__________ JOB NO.__________

DISCIPLINE__________

PREPARED BY__________

| ASSIGNED HRS. | | | | % COMPLETE | ASSN. HR x % | | | SQ CH. ISSUE / RETURN | DESIGN CHG.HR | OTHER HRS. | REMARKS | EST.HRS. TO COMP. |
|---|---|---|---|---|---|---|---|---|---|---|---|---|
| TOTAL | DE | DR | CK | 20 40 60 80 | DE | DR | CK | | | | | |
| | | | | | | | | | | | PAGE TOTALS | |

FIGURE 1.8. Typical Drawing Register and Status (Courtesy of S. I. P. Engineering, Inc., Houston, Texas)

design, drafting, and checking phases can be evaluated, and the status of individual drawings evaluated.

When assigning hours and percents completed, and when making the calculations, remember that the assigned hours are estimated, the percent completed is estimated, and the assigned hours times percent completed is the multiplication of two estimated numbers. The resulting multiplication should be *only whole* numbers.

# TYPES OF ELECTRICAL DRAWINGS

# 2

## 2.1 BASIC CATEGORIES

Electrical drawings generally fall into four broad categories:

1. Plans
2. Schedules
3. Diagrams
4. Details and/or elevations

We will examine each of these categories briefly and determine their place in the overall picture or project.

## 2.2 PLANS AND PLAN VIEWS

### 2.2.1 Definition

A plan or plan view of an object is a view seen by the observer as if he or she were looking directly down at the object. It is a two-dimensional view showing width and length. In this broad category are a number of different types of plans. The plan may be only a small portion of a plant or building or it may cover a large area, depending on the scale used.

Sometimes it is necessary to draw an area that is extremely large and, using a scale where the different items in this area have some distinction, it will be impossible to show it all on one sheet of paper. In this case, several sheets or as many as necessary can be used and a *match line* is drawn to show the plane at which the area stops on one sheet and starts on another. Notations on the match line show what drawing the plan is continued on and usually a coordinate for reference. Figures 2.1 through 2.3 show match lines in use.

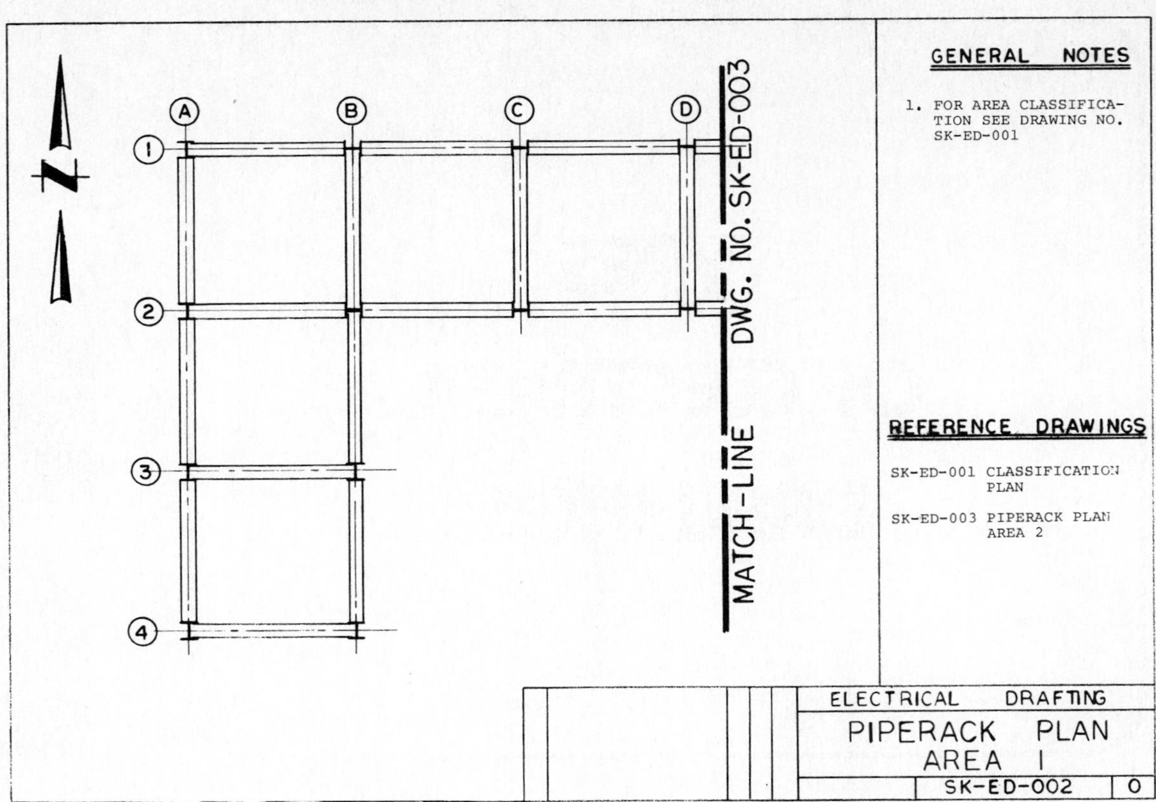


**FIGURE 2.1.** Use of Match Lines

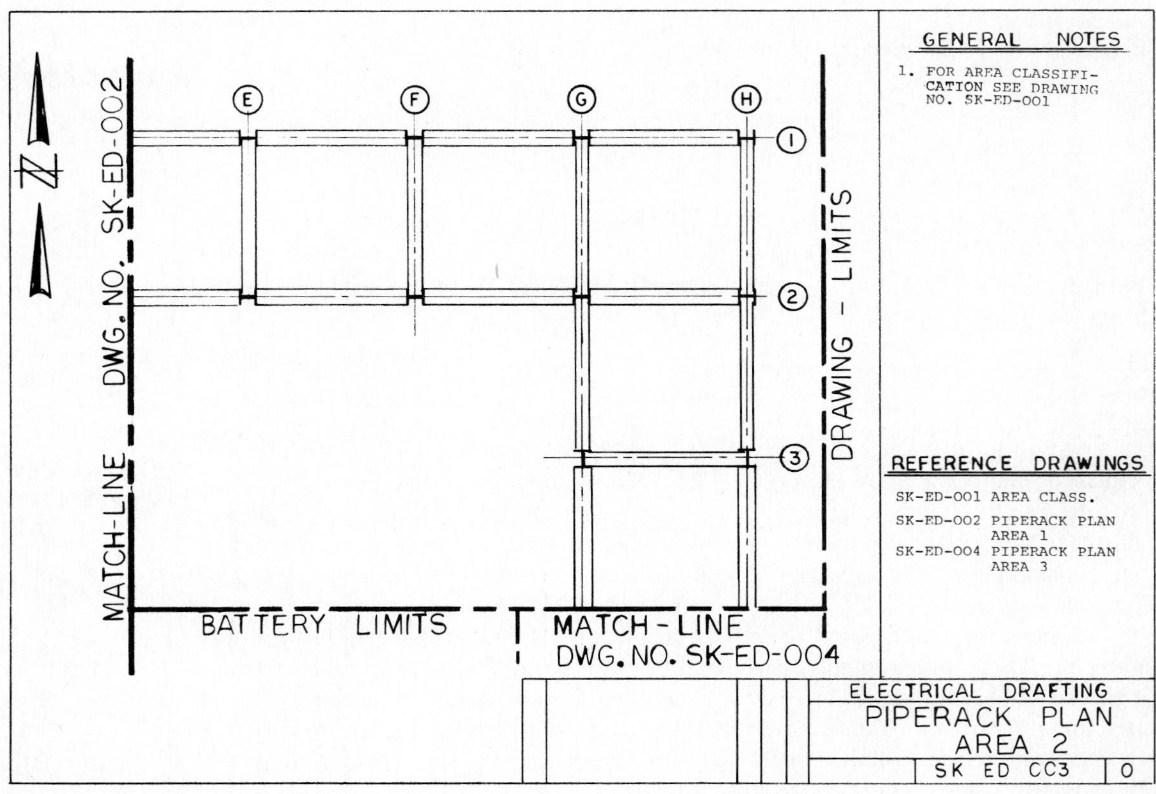


**FIGURE 2.2.** Use of Match Lines

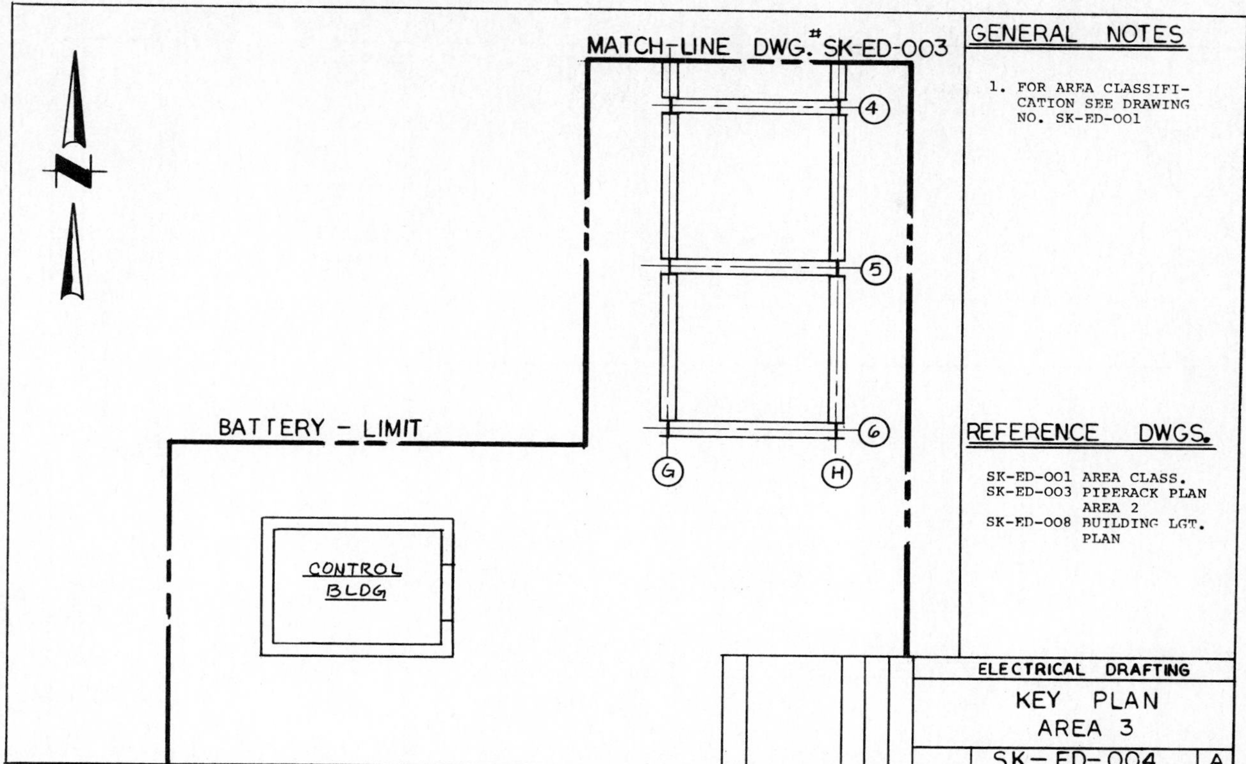


**FIGURE 2.3.** Use of Match Lines

Now let's take a look at some of the different types of plans drawn by electrical drafters.

### 2.2.2 Key Plans

Key plans show major equipment, roads, buildings, vessels, and the like in an area and list reference drawings where larger-scale or partial plan views and detail can be found. Key plans do not usually show a lot of minute detail of the area. See Figure 10.3.

### 2.2.3 Classification Plans

Some areas contain explosive or flammable materials, and extra care must be taken so that the right equipment will be installed in the area. These areas are called *classified* areas and are rated or designated by the ignition temperature or flash point of the atmosphere in the area. This is covered in depth in Chapter 11. A classification plan, by the use of symbols, shows those areas that are dangerous. Figure 2.4 is a typical classification plan.

### 2.2.4 Other Types of Plans

1. Power plan
2. Lighting plan
3. Instrument plan
4. Underground plan
5. Equipment plan

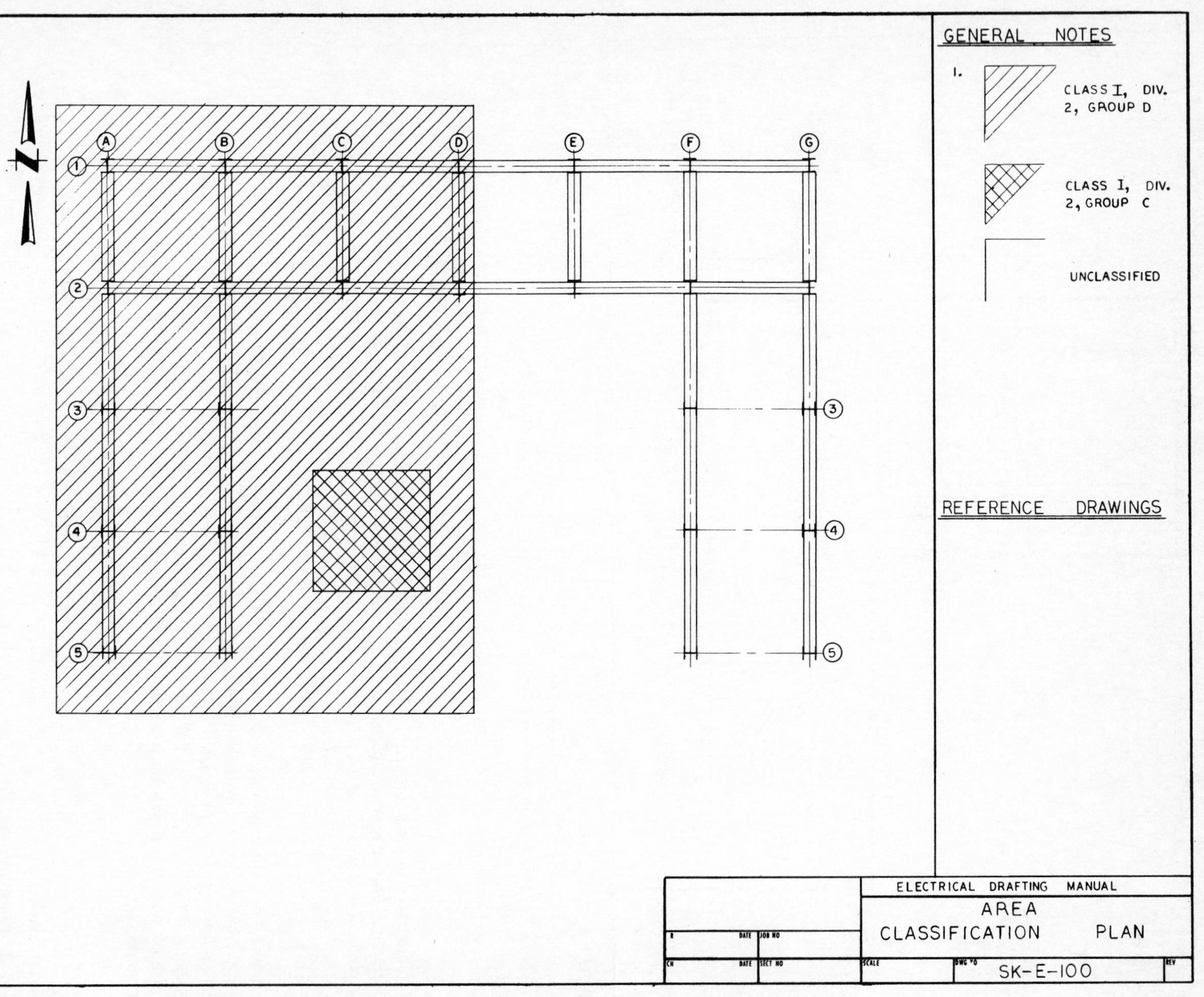


FIGURE 2.4. Typical Classification Plan

**SELF-CHECK QUIZ 2.1** **Cover the right side of the page and answer the questions.**

| | |
|---|---|
| 2.1 Electrical drawings generally fall into __________broad categories. | 2.1 Ans.: *4*<br>Ref. 2.1 |
| 2.2 A plan view is a view of an object looking directly __________at the object. | 2.2 Ans.: *down*<br>Ref. 2.2.1 |
| 2.3 Key plans do not usually show a lot of __________ __________. | 2.3 Ans.: *minute detail*<br>Ref. 2.2.2 |
| 2.4 There should be some semblance of __________in a schedule. | 2.4 Ans.: *organization*<br>Ref. 2.3.1 |
| 2.5 Four categories of drawings are __________, __________, __________, and __________. | 2.5 Ans.: *plans, schedules, diagrams,* and *details*<br>Ref. 2.1 |
| 2.6 A listing of all conduit runs on a project is called a __________and __________ schedule. | 2.6 Ans.: *conduit* and *cable*<br>Ref. 2.3.2 |
| 2.7 Detail sheets are usually limited to __________category per sheet. | 2.7 Ans.: *one*<br>Ref. 2.5.2 |
| 2.8 Schedules generally present a __________or __________of information. | 2.8 Ans.: *summary* or *collection*<br>Ref. 2.3.1 |

6. Demolition plan
7. Heat-trace plan
8. Grounding plan
9. Pole-line plan
10. Conduit stub-up plan
11. Partial plan
12. Building plan
13. Conduit plan

## 2.3 SCHEDULES

### 2.3.1 Definition

Schedules are generally full-sized drawings presenting a summary or collection of information. They are usually in tabular form. There should be some semblance of organization in a schedule, be it numerical, alphabetical or otherwise, so that the information is retrievable without scrutinizing every item in the schedule. The information should be complete with reference drawings, legends, or notes to make it self-explanatory. Some examples of schedules are described next.

### 2.3.2 Conduit and Cable Schedule

This schedule is a numerical listing of all conduit or cable runs on a particular project or in an area, giving all pertinent data such as size, type, wire fill, length, origin, destination, voltage and purpose. See Figure 10.9.

| MOTOR AND FEEDER SCHEDULE | | | | | | | |
|---|---|---|---|---|---|---|---|
| MOTOR NO. | HP | FLC | SERVICE | | SCHEMATIC DWG. NO. | LOCATION DWG. NO. | REMARKS |
| | | | WIRE | FROM | | | |
| | | | | | | | |
| | | | | | | | |
| | | | | | | | |
| | | | | | | | |
| | | | | | | | |
| | | | | | | | |
| | | | | | | | |
| | | | | | | | |
| | | | | | | | |

**FIGURE 2.5.** Motor and Feeder Schedule

### 2.3.3 Motor and Feeder Schedule

This schedule is a listing by motor number of all motors on a particular project or in an area, showing horsepower, kilowatt rating, voltage, conduit, wire, service, purpose, and control station location. See Figure 2.5.

### 2.3.4 Other Types of Schedules

1. Junction box schedule
2. Transformer schedule
3. Fixture or device schedule
4. MCC schedule
5. Switchrack schedule
6. Heat-trace schedule
7. Multiconductor cable schedule

Some or all of these schedules may be part of a project package.

## 2.4 DIAGRAMS

### 2.4.1 General

Diagrams show the electrical path, device wiring, sequence of operation, device relationship, or connections and hookups of the electrical installation. Definitions of the various types of diagrams will be given later.

### 2.4.2 Types of Diagrams

1. One-line diagram
2. Ladder diagram
3. Logic diagram
4. Schematic wiring diagram
5. Panel wiring diagram
6. Miscellaneous wiring diagrams
7. Field wiring diagram
8. Instrument wiring diagram
9. Interconnection diagram
10. External connection diagram

## 2.5 DETAIL DRAWINGS

### 2.5.1 Definition

Detail drawings, or sheets, are full-sized drawings showing complete information on a specific installation, conduit routing, equipment placement or connection, and so on. They may be elevation or plan views and are generally to a larger scale so that more detail can be called out. Several details, sometimes as many as nine, appear on one sheet.

### 2.5.2 Types of Detail Drawings

Detail sheets are usually limited to one category of details per sheet, such as power details, lighting details, or grounding details. See Figure 10.18.

Some types of detail drawings are the following:

1. Power details
2. Lighting details
3. Grounding details
4. Instrument details
5. Miscellaneous details
6. Heat-trace details
7. Standard details
8. Pole-line details
9. Motor connection details
10. Switchrack details
11. Underground conduit details
12. Elevations

# 3 BASIC CALCULATIONS

## 3.1 GENERAL

A working knowledge of the basic calculations used in electrical drafting is beneficial to the drafter. Although the responsibility for such calculations rests with the design group, a familiarity with the methods used will provide a means to better understand the intricacies of the job being done.

In this section we will look at the basic formulas used in the following:

1. Direct-current circuits
2. Sizing conduit and calculating wire fill
3. Sizing wire and determining ampacities

## 3.2 DIRECT-CURRENT CIRCUITS

We start with some definitions:

*Direct current (DC):* A direct-current circuit is one in which the flow of current is in one and only one direction (see Figure 3.1).

*Current flow (intensity), I:* The movement of electrons along a wire or conductor expressed in amperes.

*Voltage or electromotive force, E:* The force that causes the flow of electrons along a wire or conductor, expressed in volts.

*Resistance, R:* That characteristic of a circuit, always present, that tends to prevent or restrict the flow of electrons, expressed in ohms.

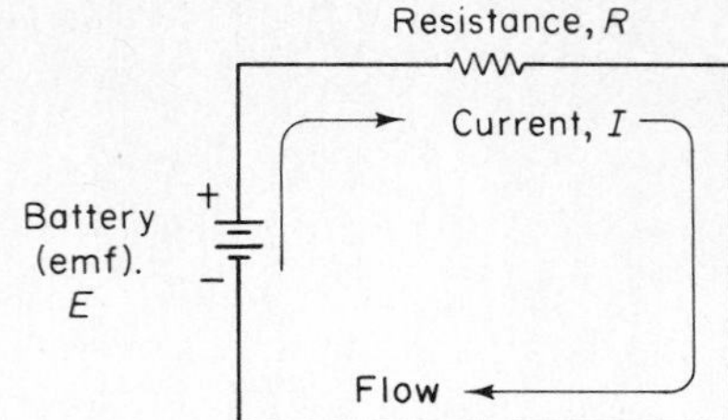


**FIGURE 3.1.** Simple Direct-Current Circuit

We can use a water analogy to explain the preceding terms:

1. If a garden hose is considered the conductor or path for water as a piece of wire is the conductor or path for the electrons to move along, the movement of water in the hose is analogous to the flow of electrons, or current, in the wire. The unit of current flow is the ampere or AMP (A).
2. Just as there is no flow of water in the hose if it is not connected to a faucet, there will be no flow of electrons in a wire unless the wire is connected to a power source such as a battery. The pressure or force that causes the electrons to flow is called electromotive force (EMF) or voltage and is expressed in volts (V).
3. The friction of the water in the hose, any kinks, knots, squeezed together sections, or other restriction will tend to slow down the flow of water. This action is similar to the resistance in an electrical circuit, which tends to slow down or restrict the flow of electrons along a conductor. The unit of resistance is the ohm (Ω).
4. Unlike the electric circuit, water can run out the end of the hose. The electrons must have a completed circuit before they can move, a path for them to return to the device from whence they came.

Direct-current circuits fall into three categories: series circuits, parallel circuits, and series–parallel or combination circuits.

### 3.2.1 Direct-Current Series Circuits

In a series circuit, all resistances are in series or tied end to end. The current must flow through each individually to complete its circuit (see Figure 3.2).

A German physicist, Gustav Kirchhoff, while experimenting with direct current, developed the law that states:

> In any direct-current circuit in which all resistances are connected in series, these resistances may be replaced by one equivalent resistance without changing the current of the circuit.

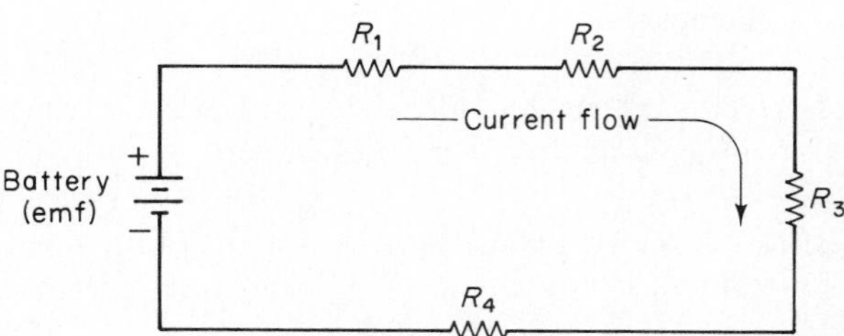


**FIGURE 3.2.** Series DC Circuit

The formula to determine the value of this equivalent resistance, $R_E$, in a series circuit is

$$R_E = R_1 + R_2 + R_3 + R_4 + \cdots$$

Or, simply, the equivalent resistance is equal to the sum of all the individual resistances in the circuit. A word of caution: This applies *only* to series circuits.

To utilize this formula, we will assign numerical values to the resistances in the circuit in Figure 3.2.

$$R_1 = 8\ \Omega, \quad R_2 = 2\ \Omega, \quad R_3 = 5\ \Omega, \quad R_4 = 5\ \Omega$$

Calculating the equivalent resistance,

$$R_E = R_1 + R_2 + R_3 + R_4$$

Substituting

$$R_E = 8 + 2 + 5 + 5 = 20\ \Omega$$

We now know that the four resistances $R_1$, $R_2$, $R_3$, and $R_4$ can be replaced with a single resistance $R_E$ valued at 20 Ω, and it will not change or alter the overall current of the circuit. The series circuit in Figure 3.2 can now be drawn as shown in Figure 3.3.

With the circuit simplified to a single resistance, and knowing the value of the battery, 100 V, and the equivalent resistance $R_E$, 20 Ω, we can determine the current flow in the circuit using Ohm's law. Georg Simon Ohm developed a relationship of current, voltage, and resistance in a DC circuit that states:

> The voltage in a direct-current circuit, or an alternating-current circuit with pure resistance, is directly proportional to the product of the current value times the resistance value.

Expressed as a formula,

$$E = IR$$

where $E$ = emf or voltage in volts
$I$ = current flow in amperes
$R$ = resistance in ohms

From this we can derive formulas to find the following:

1. Voltage: $E = IR$
2. Current: $I = E/R$
3. Resistance: $R = E/I$

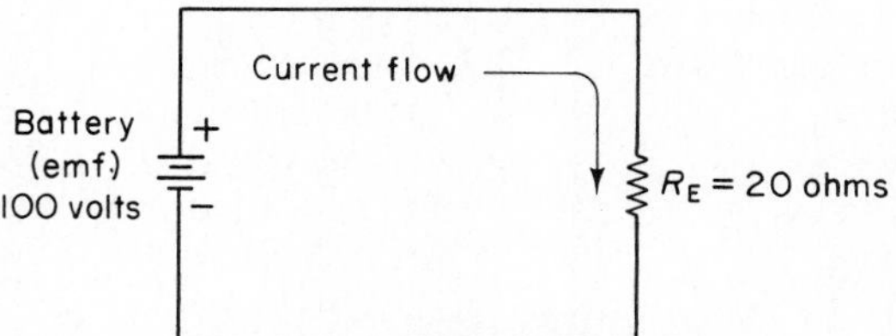


**FIGURE 3.3.** Series Circuit with Equivalent Resistance

**SELF-CHECK QUIZ 3.1** **Cover the right side of the page and answer the questions.**

| | |
|---|---|
| 3.1 Direct current flows in __________ and only __________ direction. | 3.1 Ans.: *one* and only *one* Ref. 3.2 |
| 3.2 The movement of electrons along a wire or conductor is called __________ __________. | 3.2 Ans.: *current flow* Ref. 3.2 |
| 3.3 The letter symbol for current is __________. | 3.3 Ans.: *I* Ref. 3.2 |
| 3.4 Voltage or EMF is expressed in __________. | 3.4 Ans.: *volts* Ref. 3.2 |
| 3.5 Resistance is expressed in __________. | 3.5 Ans.: *ohms* Ref. 3.2 |
| 3.6 Resistance tends to __________ the flow of electrons. | 3.6 Ans.: *restrict* Ref. 3.2 |
| 3.7 Resistance can be compared to __________ in a water hose. | 3.7 Ans.: *friction* Ref. 3.2 |
| 3.8 Three types of direct-current circuits are __________, __________, and __________–__________ or __________. | 3.8 Ans.: *series, parallel,* and *series–parallel* or *combination* Ref. 3.2 |

Since the current, $I$, is the unknown quantity in the circuit in Figure 3.3, formula 2 would be used: $I = E/R$. Substituting our known values of $E = 100$ V and $R = 20\ \Omega$, we have

$$I = \frac{E}{R} = \frac{100}{20} = 5 \text{ A}$$

If any two values of a direct-current circuit are known, the third can be found by using Ohm's law.

***Summary: Solving DC Series Circuits.***

1. Sketch the circuit showing all values known.
2. Simplify the circuit by applying Kirchhoff's law, $R_E = R_1 + R_2 + R_3 + \cdots$ until the circuit contains only one resistance, $R_E$.
3. Redraw the circuit with $R_E$.
4. Substitute the values in Ohm's law to find the unknown.

### 3.2.2 Direct-Current Parallel Circuits

In dc parallel circuits, resistances are connected side by side, as opposed to end to end in the series circuit (see Figure 3.4). Kirchhoff also determined that in a parallel dc circuit an equivalent resistance can be substituted for all parallel resistances by using the formula

$$\frac{1}{R_E} = \frac{1}{R_1} + \frac{1}{R_2} + \frac{1}{R_3} + \cdots$$

Once again, let's assign values to our resistances:

$$R_1 = 10\ \Omega, \quad R_2 = 5\ \Omega, \quad R_3 = 10\ \Omega$$

The battery voltage is 100 V.

*Note:* The battery voltage will be applied simultaneously across each resistance in the circuit.

Substitutions in our formula will yield

$$\frac{1}{R_E} = \frac{1}{R_1} + \frac{1}{R_2} + \frac{1}{R_3}$$
$$= {}^{1}/_{10} + {}^{1}/_{5} + {}^{1}/_{10}$$

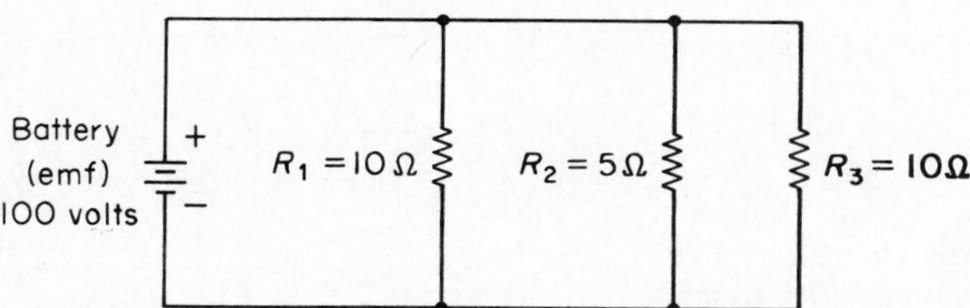


**FIGURE 3.4.** Parallel DC Circuit

We determine a common denominator of 10 and proceed:

$$\frac{1}{R_E} = {}^1/_{10} + {}^2/_{10} + {}^1/_{10} = {}^4/_{10}$$

We can invert both sides of our equation without changing the value and get

$$\frac{R_E}{1} = {}^{10}/_4 \quad \text{or} \quad R_E = {}^{10}/_4 = 2.5\ \Omega$$

*Note:* The equivalent resistance will always be smaller than the value of any of the individual resistances.

Our circuit now looks like Figure 3.5.

Using Ohm's law again, we can substitute our two known values of voltage ($E$) and resistance ($R$) to solve for current:

$$I = \frac{E}{R} = \frac{100}{2.5} = 40\ \text{A}$$

*Note:* If the resistances in a parallel DC circuit all have the *same value,* the equivalent resistance is the value of one resistance divided by the number of resistances in the circuit.

***Summary for Solving DC Parallel Circuits.*** To apply Ohm's law to a dc parallel circuit,

1. Draw on sketch the circuit showing all values known.
2. Simplify the circuit by applying Kirchhoff's law for parallel circuits to all or parts of the circuit: $1/R_E = 1/R_1 + 1/R_2 + 1/R_3 + \cdots$
3. Redraw the circuit with substitutions. The value of $R_E$ is substituted for all the other resistances.
4. Substitute the values for $E$ (100 V) and $R$ (2.5 Ω) in the Ohm's law formula: $I = E/R = 100/2.5 = 40$ A (total current).

### 3.2.3 Direct-Current Series–Parallel Circuits

A dc series–parallel circuit, sometimes called a combination circuit, contains resistances in series, as well as resistances in parallel in the same circuit (see Figure 3.6).

***Summary for Solving DC Series–Parallel Circuits.***

1. Simplification of a series–parallel circuit is a step-by-step application of Kirchhoff's laws for series circuits and for parallel circuits.

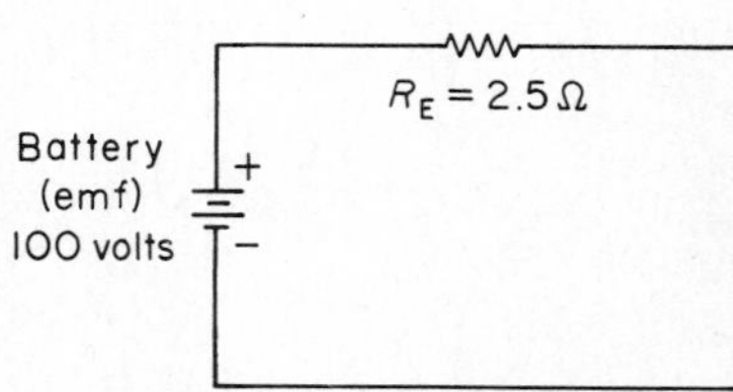


**FIGURE 3.5.** Equivalent Resistance Circuit

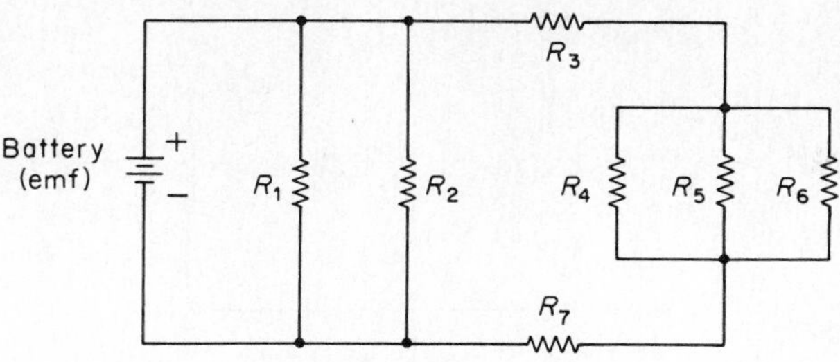


**FIGURE 3.6.** Series–Parallel or Combination Circuit

2. Let's assign values to each resistor:

$$R_1 = 1, \quad R_2 = 2, \quad R_3 = 3, \quad R_4 = 4$$
$$R_5 = 5, \quad R_6 = 6, \quad R_7 = 7, \quad E = 100 \text{ V}$$

3. Looking at Figure 3.6, we see that $R_4$, $R_5$, and $R_6$ are in parallel and can be replaced by an equivalent resistance as follows:

$$\frac{1}{R_{E_1}} = \frac{1}{R_4} + \frac{1}{R_5} + \frac{1}{R_6}$$

Substitutions of values in our formula results in

$$\frac{1}{R_{E_1}} = {}^1/_4 + {}^1/_5 + {}^1/_6 = {}^{15}/_{60} + {}^{12}/_{60} + {}^{10}/_{60}$$

So

$$\frac{1}{R_{E_1}} = {}^{37}/_{60} \quad \text{or} \quad \frac{R_E}{1} = {}^{60}/_{37}$$

$$R_{E_1} = 1.62\ \Omega$$

Redrawing our circuit, it now looks like Figure 3.7.

4. We now have $R_3$ and $R_7$ in series with $R_{E_1}$, so using the series-equivalent resistance formula

$$R_{E_2} = R_3 + R_{E_1} + R_7$$

We substitute our values and our formula reads

$$R_{E_2} = 3 + 1.62 + 7 = 11.62$$

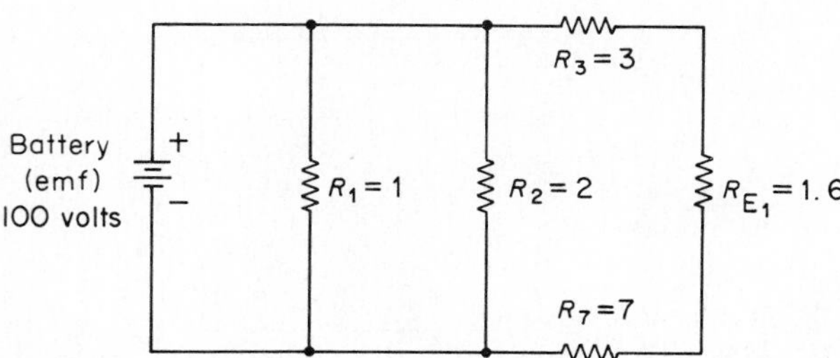


**FIGURE 3.7.** Series–Parallel Circuit Partially Simplified

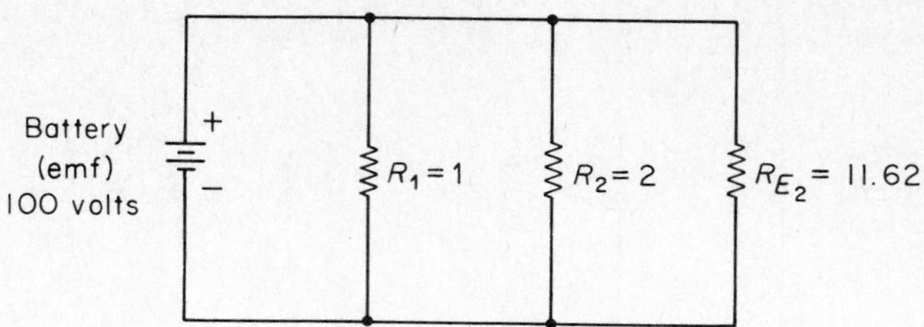


**FIGURE 3.8.** Parallel Circuit

Now our circuit has been simplified as shown in Figure 3.8. We now have three resistances in parallel, so

$$\frac{1}{R_{ET}} = \frac{1}{R_1} + \frac{1}{R_2} + \frac{1}{R_{E_2}} = \frac{1}{1} + \frac{1}{2} + \frac{1}{11.62}$$

$$= \frac{11.62}{11.62} + \frac{5.81}{11.62} + \frac{1}{11.62} = \frac{18.43}{11.62}$$

or

$$\frac{R}{ET} = \frac{11.62}{18.43} = 0.63\ \Omega$$

We now have the circuit shown in Figure 3.9, and we can apply Ohm's law, $E = IR$, and solve for the current. Rearranging the formula to solve for current ($I$), we have $I = E/R$, and substituting our known values of $E = 100$ V and $R = 0.63$, we have $I = 100/0.63 = 158.7$ A, which is the total current flow in our circuit. Thus, in a series–parallel circuit the resistances can be grouped into series or parallel resistances and replaced. Then we group again and replace, until we have replaced the complicated circuit with one having only one resistance. Then we can apply Ohm's law and solve for the current.

## 3.3 ALTERNATING CURRENT

### 3.3.1 General

At least 90 percent of all electricity consumed by industry throughout the world is alternating current (ac). Our homes and factories are powered by alternating current because it is much more economical to generate and distribute than dc. The main reason for this is that the power lost during the transmission of ac from the generating station to the user is very much less than with dc. Using ac, the power companies are able to *transform* the produced electrical energy into a high-voltage, low-current equivalent power with a device called a *transformer* working on the principle of mutual induction.

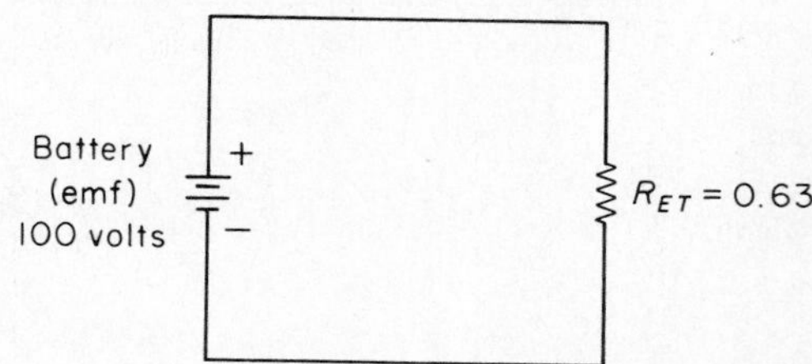


**FIGURE 3.9.** Simplified DC Circuit

### 3.3.2 Summary of Terms Used

A basic understanding of alternating current requires the use of a number of general terms and laws. When applied to electrical circuits, these general terms assume a very specific meaning.

***Definitions***

*Alternating Current:* Electric current that continually changes directions in a circuit as compared to direct current, which flows in one and only one direction.

*Potential:* The amount of charge held by one body as compared to another point or body. It is usually measured in volts.

*Cycle:* A unit of measure completed when an alternating current has passed through a complete set of values in both directions; when 360 electrical degrees of a sine wave have been completed.

*Effective value:* That value of an alternating current that will produce the same heating effect as the same value of direct current. The formula for effective value is effective value = 7.07 maximum peak value.

*Unit of frequency:* The hertz; one hertz (Hz) is equal to one cycle per second. Commercial power in the United States is 60 Hz or 60 cycles per second. In many foreign countries a frequency of 50 Hz is used.

*Induction:* The act or process of producing voltage by the relative motion of a conductor across a magnetic field. This is the basic principle of the electric generator or alternator. When implemented inversely so that voltage is applied to cause motion of a conductor, it becomes the principle of the electric motor.

*Transformer action:* A method of transferring electrical energy from one coil or core to another by an alternating magnetic field, sometimes called *mutual induction*. The coil that generates the magnetic field is called the *primary* side of the transformer and the coil in which the voltage is induced is called the *secondary*. The value of the voltage induced in the secondary can be stepped up or stepped down depending on the ratio of the turns in the secondary to the turns in the primary.

$$V_{sec} = \frac{N_{sec}}{N_{pri}} \times V_{pri}$$

***Laws.***

*Ohm's law:* As we observed in the discussion of direct current, Ohm's law expresses the relationship that exists between voltage, current, and resistance. Ohm's law formulas can also be used in ac circuits.

*Joule's law:* The flow of current through a resistance is accompanied by heat. Joule's experiments showed that the electrical energy (in watts) dissipated in a metallic wire is the product of the square of the current and the time during which current flows.

$$W = RI^2T$$

This equation is known as Joule's law. $I$ is in amperes, $T$ in seconds, $W$ in Joules or watt-seconds, and $R$ in ohms. This equation gives another definition of resistance

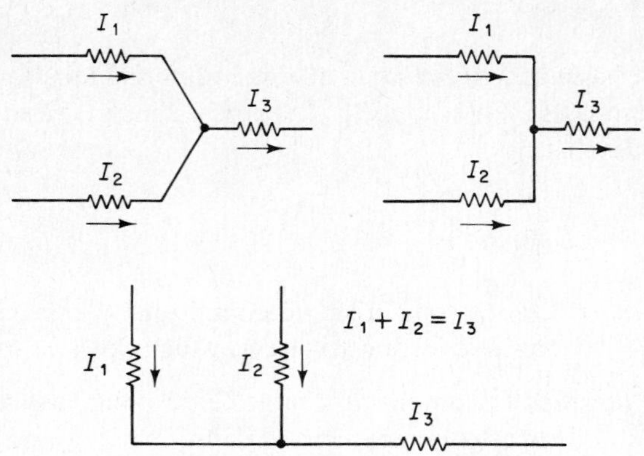


**FIGURE 3.10.** Kirchhoff's Current Law

($R$) from an energy point of view: resistance is that element of a circuit which accounts for the existence of heat. Without heat from resistance, you would not have electric toasters, stoves, heaters, incandescent lamps, blankets, and many other electrical servants.

*Kirchhoff s laws:* Complex electric circuits can be solved with the aid of two simple rules known as Kirchhoff's laws.

*Kirchhoff s current law:* The sum of the current flowing away from any point in an electric circuit must be equal to the sum of the current flowing toward the point. The current law is the principle involved in the current rule for parallel circuits. The sum of the branch individual currents must equal the total line currents (Figure 3.10).

*Kirchhoff s voltage law:* Around any closed path in an electric circuit, the sum of the voltage drops is equal to the sum of the applied voltage (Figure 3.11).

By means of a systematic application of these laws, the current and voltage value of the various parts of a complex circuit can be found.

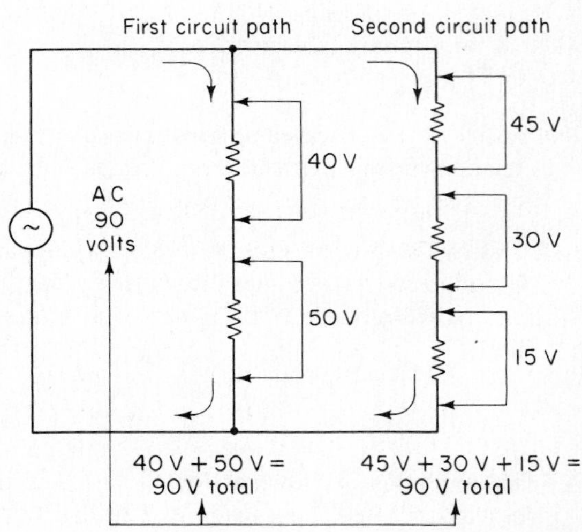


**FIGURE 3.11.** Kirchhoff's Voltage Law

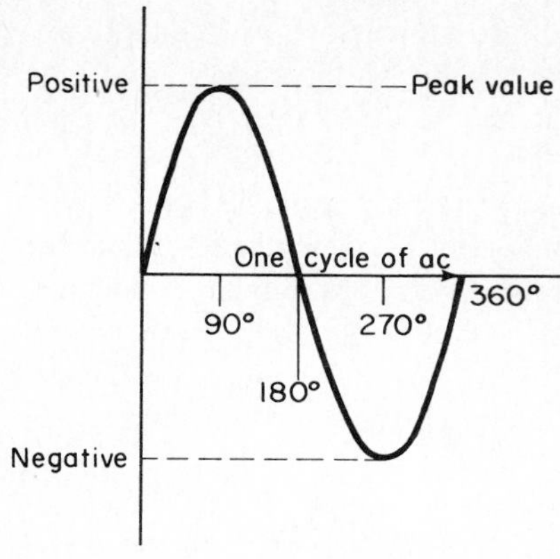


**FIGURE 3.12.** Sine Wave

### 3.3.3 Wave Forms

Wave forms are pictures that show how current and voltage change over a period of time. The value of voltage or current is represented in the vertical direction, while time is represented on the horizontal line. Wave forms for direct current are straight lines; neither the voltage nor current varies with time.

### 3.3.4 Alternating-Current Cycle

An alternating voltage or current rises from zero to a maximum value, 90°, then decreases back to zero, 180°. It increases to a maximum value in the opposite direction, 270°, and then decreases back to zero, 360°. The complete set of values passed through is called a *cycle,* as illustrated in Figure 3.12. The cycle is repeated as long as current flows.

The wave forms of alternating current are curved to represent gradual changes in voltage and current, first increasing, then decreasing in value for each direction of current flow. Most alternating current that we work with has a wave form represented by a sine curve. While alternating currents and voltages do not always have wave forms of exact sine curves, they are usually assumed to have a sine wave form.

### 3.3.5 Summary

An in-depth understanding of alternating current and all its ramifications involves years of study and experience. This short presentation is merely to make you aware of its existence and some of the terms used.

## 3.4 CONDUCTOR SPECIFICATIONS

### 3.4.1 General

Wires and cables are manufactured in a great many different types and forms of construction. These various types and forms have been developed over the years to meet the demand for installation under all manner of circumstances. Cables that run in ducts under the street must be able to function when completely under water because many times the duct lines are flooded. Wires that serve controls and other devices needed for operation of ovens and furnaces must be able to withstand high temperatures. Cables supplying power to electric mining equipment must have tough jackets and insulations to take the abuse from abrasion against rocks as they are dragged across the ground. The proper choice of a wire or cable depends on the consideration of many factors to make sure it will perform safely.

Annealed copper is the material generally used as a conductor in insulated wires and cables. When covered with rubber insulation, the copper is sometimes coated with tin or a lead alloy for protection of both the insulation and the copper because each contains substances that attack the other.

Aluminum is also used as a material for conducting electricity. The conductivity of aluminum is 61 percent that of copper; therefore, to carry the same current as copper conductors, aluminum conductors must be considerably larger. If aluminum conductors are large enough in circumference to have the same conductivity as a given copper conductor, their weight per unit length is still only about half that of the copper conductor. Many long-span, high-voltage distribution lines use aluminum wire due to this weight factor.

The American wire gauge (AWG) is used exclusively in the United States for copper, aluminum, and bronze wire. Wires and cables are sized according to the AWG from small sizes such as #18, #16, #14 up through #0000 (commonly written 4/0 and called four aught). From 4/0 up through all the larger sizes, cables are rated for size in circular mils. This is basically an expression in terms of total cross-sectional area of the conductor. The area of a circle, in circular mils, is equal to the square of its diameter in mils. For example, a wire with a 10-mil diameter would be rated for wire size as 100-circular mil wire (10 mils × 10 mils = 100 circular mils). Mil values are calculated by multiplying diameter in inches × 1000. To convert diameter to inches from mils, divide by 1000.

The standard wire sizes listed in tables in the NEC (National Electric Code) in AWG and circular mils are given in the table of Figure 3.15. The abbreviation MCM is used for circular mil cables with the first letter M signifying 1000. As an example, 250 MCM has a cross-sectional area of 250,000 circular mils.

The National Electric Code lists the many types of conductors that are used for general wiring. Each type has been given a letter designation, and these wires and cables are often called *code grade*. The code provides a table for conductor applications and a table for conductor insulations. A majority of wire and cable installations use the code grade conductors. These may be found in Section 310-3, pages 131 through 138, in the 1981 edition of the National Electric Code.

To provide uniformity and to enable us to be accurate in sizing of conductors, the NEC publishes tables of ampacities for the different sizes of wire and types of insulation under different physical conditions. Two of these tables appear in Figure 3.13. They may also be found in Chapter 9 of your codebook.

### 3.4.2 Ampacities

Ampacity is another way of saying current-carrying capacity and generally refers to wire and wire sizes. When determining the wire size necessary to operate a specific load (motor, lamps, etc.), several factors must be known: (1) the voltage of the device, and (2) the current requirements and limitations of the device.

The wire selected must be of sufficient cross section to meet the maximum current requirements of the load and at the same time the temperature of the wire must remain below the rated temperature of the wire insulation. If the wire is too small, problems such as overheating will occur. When excess heat is dissipated, it represents a loss in power and additional energy costs. Overheating, if excessive, can damage the insulation, resulting in faults in the circuit, short or grounded circuits, and possible fire.

By using the ampacity tables in Figure 3.13, it is possible to determine the following:

1. The wire size necessary to carry a specified current if the current in amperes is known and the type of insulation is known.

**Table 310-16. Allowable Ampacities of Insulated Conductors Rated 0-2000 Volts, 60° to 90°C**

Not More Than Three Conductors in Raceway or Cable or Earth (Directly Buried), Based on Ambient Temperature of 30°C (86°F)

| Size | Temperature Rating of Conductor, See Table 310-13 | | | | | | | | Size |
|---|---|---|---|---|---|---|---|---|---|
| | 60°C (140°F) | 75°C (167°F) | 85°C (185°F) | 90°C (194°F) | 60°C (140°F) | 75°C (167°F) | 85°C (185°F) | 90°C (194°F) | |
| AWG MCM | TYPES †RUW, †T, †TW, †UF | TYPES †FEPW, †RH, †RHW, †RUH, †THW, †THWN, †XHHW, †USE, †ZW | TYPES V, MI | TYPES TA, TBS, SA, AVB, SIS, †FEP, †FEPB, †RHH †THHN, †XHHW* | TYPES †RUW, †T, †TW, †UF | TYPES †RH, †RHW, †RUH, †THW †THWN, †XHHW, †USE | TYPES V, MI | TYPES TA, TBS, SA, AVB, SIS, †RHH, †THHN, †XHHW* | AWG MCM |
| | COPPER | | | | ALUMINUM OR COPPER-CLAD ALUMINUM | | | | |
| 18 | .... | .... | .... | 14 | .... | .... | .... | .... | .... |
| 16 | .... | .... | 18 | 18 | .... | .... | .... | .... | .... |
| 14 | 20† | 20† | 25 | 25† | .... | .... | .... | .... | .... |
| 12 | 25† | 25† | 30 | 30† | 20† | 20† | 25 | 25† | 12 |
| 10 | 30† | 35† | 40 | 40† | 25† | 30† | 30 | 35† | 10 |
| 8 | 40 | 50 | 55 | 55 | 30 | 40 | 40 | 45 | 8 |
| 6 | 55 | 65 | 70 | 75 | 40 | 50 | 55 | 60 | 6 |
| 4 | 70 | 85 | 95 | 95 | 55 | 65 | 75 | 75 | 4 |
| 3 | 85 | 100 | 110 | 110 | 65 | 75 | 85 | 85 | 3 |
| 2 | 95 | 115 | 125 | 130 | 75 | 90 | 100 | 100 | 2 |
| 1 | 110 | 130 | 145 | 150 | 85 | 100 | 110 | 115 | 1 |
| 0 | 125 | 150 | 165 | 170 | 100 | 120 | 130 | 135 | 0 |
| 00 | 145 | 175 | 190 | 195 | 115 | 135 | 145 | 150 | 00 |
| 000 | 165 | 200 | 215 | 225 | 130 | 155 | 170 | 175 | 000 |
| 0000 | 195 | 230 | 250 | 260 | 150 | 180 | 195 | 205 | 0000 |
| 250 | 215 | 255 | 275 | 290 | 170 | 205 | 220 | 230 | 250 |
| 300 | 240 | 285 | 310 | 320 | 190 | 230 | 250 | 255 | 300 |
| 350 | 260 | 310 | 340 | 350 | 210 | 250 | 270 | 280 | 350 |
| 400 | 280 | 335 | 365 | 380 | 225 | 270 | 295 | 305 | 400 |
| 500 | 320 | 380 | 415 | 430 | 260 | 310 | 335 | 350 | 500 |
| 600 | 355 | 420 | 460 | 475 | 285 | 340 | 370 | 385 | 600 |
| 700 | 385 | 460 | 500 | 520 | 310 | 375 | 405 | 420 | 700 |
| 750 | 400 | 475 | 515 | 535 | 320 | 385 | 420 | 435 | 750 |
| 800 | 410 | 490 | 535 | 555 | 330 | 395 | 430 | 450 | 800 |
| 900 | 435 | 520 | 565 | 585 | 355 | 425 | 465 | 480 | 900 |
| 1000 | 455 | 545 | 590 | 615 | 375 | 445 | 485 | 500 | 1000 |
| 1250 | 495 | 590 | 640 | 665 | 405 | 485 | 525 | 545 | 1250 |
| 1500 | 520 | 625 | 680 | 705 | 435 | 520 | 565 | 585 | 1500 |
| 1750 | 545 | 650 | 705 | 735 | 455 | 545 | 595 | 615 | 1750 |
| 2000 | 560 | 665 | 725 | 750 | 470 | 560 | 610 | 630 | 2000 |
| | CORRECTION FACTORS | | | | | | | | |
| Ambient Temp. °C | For ambient temperatures over 30°C, multiply the ampacities shown above by the appropriate correction factor to determine the maximum allowable load current. | | | | | | | | Ambient Temp. °F |
| 31-40 | .82 | .88 | .90 | .91 | .82 | .88 | .90 | .91 | 86-104 |
| 41-45 | .71 | .82 | .85 | .87 | .71 | .82 | .85 | .87 | 105-113 |
| 46-50 | .58 | .75 | .80 | .82 | .58 | .75 | .80 | .82 | 114-122 |
| 51-60 | .... | .58 | .67 | .71 | .... | .58 | .67 | .71 | 123-141 |
| 61-70 | .... | .35 | .52 | .58 | .... | .35 | .52 | .58 | 142-158 |
| 71-80 | .... | .... | .30 | .41 | .... | .... | .30 | .41 | 159-176 |

† The load current rating and the overcurrent protection for conductor types marked with an obelisk (†) shall not exceed 15 amperes for 14 AWG, 20 amperes for 12 AWG, and 30 amperes for 10 AWG copper; or 15 amperes for 12 AWG and 25 amperes for 10 AWG aluminum and copper-clad aluminum.

* For dry locations only. See 75°C column for wet locations.

**FIGURE 3.13.** Ampacity Tables (Courtesy National Fire Protection Association, National Electric Code, 1981)

**Table 310-17. Allowable Ampacities of Insulated Conductors Rated 0-2000 Volts, 60° to 90°C**

Single conductors in free air, based on ambient temperature of 30°C (86°F).

| Size | Temperature Rating of Conductor, See Table 310-13 | | | | | | | | Size |
|---|---|---|---|---|---|---|---|---|---|
| | 60°C (140°F) | 75°C (167°F) | 85°C (185°F) | 90°C (194°F) | 60°C (140°F) | 75°C (167°F) | 85°C (185°F) | 90°C (194°F) | |
| AWG MCM | TYPES †RUW, †T, †TW | TYPES †FEPW, †RH, †RHW, †RUH, †THW, †THWN, †XHHW, †ZW | TYPES V, MI | TYPES TA, TBS, SA, AVB, SIS, †FEP, †FEPB, †RHH †THHN, †XHHW* | TYPES †RUW, †T, †TW | TYPES †RH, †RHW, †RUH, †THW, †THWN, †XHHW | TYPES V, MI | TYPES TA, TBS, SA, AVB, SIS, †RHH, †THHN, †XHHW* | AWG MCM |
| | COPPER | | | | ALUMINUM OR COPPER-CLAD ALUMINUM | | | | |
| 18 | .... | .... | .... | 18 | .... | .... | .... | .... | .... |
| 16 | .... | .... | 23 | 24 | .... | .... | .... | .... | .... |
| 14 | 25† | 30† | 30 | 35† | .... | .... | .... | .... | .... |
| 12 | 30† | 35† | 40 | 40† | 25† | 30† | 30 | 35† | 12 |
| 10 | 40† | 50† | 55 | 55† | 35† | 40† | 40 | 40† | 10 |
| 8 | 60 | 70 | 75 | 80 | 45 | 55 | 60 | 60 | 8 |
| 6 | 80 | 95 | 100 | 105 | 60 | 75 | 80 | 80 | 6 |
| 4 | 105 | 125 | 135 | 140 | 80 | 100 | 105 | 110 | 4 |
| 3 | 120 | 145 | 160 | 165 | 95 | 115 | 125 | 130 | 3 |
| 2 | 140 | 170 | 185 | 190 | 110 | 135 | 145 | 150 | 2 |
| 1 | 165 | 195 | 215 | 220 | 130 | 155 | 165 | 175 | 1 |
| 0 | 195 | 230 | 250 | 260 | 150 | 180 | 195 | 205 | 0 |
| 00 | 225 | 265 | 290 | 300 | 175 | 210 | 225 | 235 | 00 |
| 000 | 260 | 310 | 335 | 350 | 200 | 240 | 265 | 275 | 000 |
| 0000 | 300 | 360 | 390 | 405 | 235 | 280 | 305 | 315 | 0000 |
| 250 | 340 | 405 | 440 | 455 | 265 | 315 | 345 | 355 | 250 |
| 300 | 375 | 445 | 485 | 505 | 290 | 350 | 380 | 395 | 300 |
| 350 | 420 | 505 | 550 | 570 | 330 | 395 | 430 | 445 | 350 |
| 400 | 455 | 545 | 595 | 615 | 355 | 425 | 465 | 480 | 400 |
| 500 | 515 | 620 | 675 | 700 | 405 | 485 | 525 | 545 | 500 |
| 600 | 575 | 690 | 750 | 780 | 455 | 540 | 595 | 615 | 600 |
| 700 | 630 | 755 | 825 | 855 | 500 | 595 | 650 | 675 | 700 |
| 750 | 655 | 785 | 855 | 885 | 515 | 620 | 675 | 700 | 750 |
| 800 | 680 | 815 | 885 | 920 | 535 | 645 | 700 | 725 | 800 |
| 900 | 730 | 870 | 950 | 985 | 580 | 700 | 760 | 785 | 900 |
| 1000 | 780 | 935 | 1020 | 1055 | 625 | 750 | 815 | 845 | 1000 |
| 1250 | 890 | 1065 | 1160 | 1200 | 710 | 855 | 930 | 960 | 1250 |
| 1500 | 980 | 1175 | 1275 | 1325 | 795 | 950 | 1035 | 1075 | 1500 |
| 1750 | 1070 | 1280 | 1395 | 1445 | 875 | 1050 | 1145 | 1185 | 1750 |
| 2000 | 1155 | 1385 | 1505 | 1560 | 960 | 1150 | 1250 | 1335 | 2000 |
| | CORRECTION FACTORS | | | | | | | | |
| Ambient Temp. °C | For ambient temperatures over 30°C, multiply the ampacities shown above by the appropriate correction factor to determine the maximum allowable load current. | | | | | | | | Ambient Temp. °F |
| 31-40 | .82 | .88 | .90 | .91 | .82 | .88 | .90 | .91 | 86-104 |
| 41-45 | .71 | .82 | .85 | .87 | .71 | .82 | .85 | .87 | 105-113 |
| 46-50 | .58 | .75 | .80 | .82 | .58 | .75 | .80 | .82 | 114-122 |
| 51-60 | .... | .58 | .67 | .71 | .... | .58 | .67 | .71 | 123-141 |
| 61-70 | .... | .35 | .52 | .58 | .... | .35 | .52 | .58 | 142-158 |
| 71-80 | .... | .... | .30 | .41 | .... | .... | .30 | .41 | 159-176 |

† The load current rating and the overcurrent protection for conductor types marked with an obelisk (†) shall not exceed 20 amperes for 14 AWG, 25 amperes for 12 AWG, and 40 amperes for 10 AWG copper, or 20 amperes for 12 AWG and 30 amperes for 10 AWG aluminum and copper-clad aluminum.

* For dry locations only. See 75°C column for wet locations.

**FIGURE 3.13.** *continued*

2. The current-carrying capacity or ampacity of a size of wire if the wire size is known and the type of insulation is known.

*Note:* It is necessary to know or specify an insulation type in both cases. The ampacity of a piece of wire will vary according to the type of insulation around it.

*EXAMPLE 1:* What is the current-carrying capacity of #6 copper, type THWN, in a raceway?

Using Table 310-16 in Figure 3-13, we locate #6 under the left column headed AWG-MCM and move directly across to the right until we come to a number in the column headed THWN. That number is 65. Thus #6 wire with THWN insulation is capable of carrying 65 a without becoming overheated. It is evident from the table that different insulations on the same AWG size wire will have a marked effect on the current-carrying capacity.

*EXAMPLE 2:* A furnace requires 120 a of current. What size of wire, type THW, is necessary to operate this furnace, if the wire is direct buried?

Using Table 310-16 in Figure 3-13, locate the column headed THW (copper). Move down this column until a number is reached that equals or exceeds the 120 a required for the device. This number is 130. Move directly to the left to the column headed AWG-MCM and we see 1. This indicates a #1 AWG size wire with type THW insulation will carry the 120 a required by the furnace without overheating.

Certain other restrictions must be taken into consideration when sizing wire.

1. The ambient temperature of the area.
2. The number of conductors in a raceway or conduit.
3. If the load is continuous or not.
4. Whether the wire is copper or aluminum.

If the ambient temperature, the temperature of the air, in the area is high, it may be necessary to use a larger size of wire to reduce the heat loss.

If the number of wires or conductors in the raceway or conduit exceeds three, the National Electric Code requires the derating of the ampacity to a percent of the value shown in the table. In some cases this requires a larger size of wire. See Note 8, page 143, 1981 edition National Electric Code.

| 480 V ac 3Ø Motor Ampacity Table | | | | | |
|---|---|---|---|---|---|
| HP | FLC | 125% X FLC | HP | FLC | 125% X FLC |
| $\frac{1}{2}$ | 1 | 1.25 | 40 | 52 | 66 |
| 1 | 1.8 | 2.2 | 50 | 65 | 81 |
| 2 | 3.4 | 4.3 | 60 | 77 | 96 |
| 5 | 7.6 | 9.4 | 75 | 96 | 122 |
| 10 | 14 | 17.5 | 100 | 124 | 155 |
| 25 | 34 | 43 | 150 | 180 | 225 |
| 30 | 40 | 50 | 200 | 240 | 300 |

**FIGURE 3.14.** Motor Ampacity Calculation Table. Note: See Table 430-150, page 345, 1981 edition, National Electric Code.

**SELF-CHECK QUIZ 3.2** **Cover the right side of the page and answer the questions.**

|  |  |
|---|---|
| 3.9 Kirchhoff's law states that the equivalent resistance in a series circuit is equal to the __________ of the individual resistances. | 3.9 Ans.: *sum* Ref. 3.2.1 |
| 3.10 Ohm's law states that the voltage in a direct-current circuit is directly proportional to the __________of the current and the resistance. | 3.10 Ans.: *product* Ref. 3.2.1 |
| 3.11 Expressed as a formula, Ohm's law is __________ = __________ | 3.11 Ans.: $E = IR$ Ref. 3.2.1 |
| 3.12 Kirchhoff's formula for the equivalent resistance in a parallel circuit is ______ = ______ + ______ + ______ + ______ ···. | 3.12 Ans.: $1/R_E = 1/R_1 + 1/R_2 + 1/R_3 + \cdots$ Ref. 3.2.2 |
| 3.13 A series-parallel circuit is sometimes called a __________circuit. | 3.13 Ans.: *combination* Ref. 3.2.3 |
| 3.14 __________ __________ is the material generally used as a conductor in insulated wires and cables. | 3.14 Ans.: *annealed copper* Ref. 3.3.1 |
| 3.15 AWG stands for __________ __________ __________ | 3.15 Ans.: *American wire gauge* Ref. 3.4.1 |
| 3.16 Ampacity is another way of saying __________-__________ __________ | 3.16 Ans.: *current-carrying capacity* Ref. 3.4.2 |

When using aluminum wire instead of copper, it is necessary to use a different wire and consequently a larger wire for the same ampacity, as aluminum only has 61 percent of the conductivity of copper.

### 3.4.3 Sizing Conductors for Single Motor Loads

The National Electric Code requires that branch-circuit conductors supplying a single motor have an ampacity of not less than 125 percent of the full-load current motor rating. In the case of single multispeed motors, the highest full-load current must be used.

Preparation of a simple calculation table like Figure 3.14 is often handy in recording the ampacities and wire sizes of project motors.

## 3.5 CONDUIT

According to Webster's dictionary, conduit is a term applied to a pipe for conveying liquids. In electrical work, a conduit is a pipe or raceway used as a housing for wires. Conduit can be manufactured from several different base materials. The following is a brief description of the various types of electrical conduit used in industrial construction.

### 3.5.1 Basic Types and Materials

Rigid steel conduit is made in two types: white (galvanized) and black (enameled). The white conduit, made of rigid galvanized steel, or RGS, is galvanized inside and out and is the type that should be used where exposed to weather, when embedded in concrete, or when installed in wet locations. The black conduit is coated with black enamel and is the type commonly used for general interior wiring. Except for special conditions, no size smaller than $^1/_2$ in. ID is allowed. All types of electrical conduit are referred to by their inside diameter (ID).

Rigid-metal-conduit wiring is approved for both exposed and concealed work and for use in nearly all classes of buildings. For ordinary conditions, wiring in metal conduit is probably the best, although it is the most expensive. The advantages of metal conduit are as follows:

1. Wires can be inserted and removed.
2. It provides a good ground path.
3. It is strong enough mechanically that nails cannot be driven through it and it is not readily deformed by blows or by wheelbarrows being run over it.
4. It successfully resists the normal chemical reaction of cement when embedded in partitions or walls of concrete buildings.

***When to Use Rigid Metal Conduit Wiring.*** In general, conduit wiring should be used whenever the job will stand the cost. Rigid conduit protects the conductors it contains and provides a smooth raceway permitting ready insertion or removal.

Conduits and fittings exposed to severe corrosive influences shall be of corrosion-resistant material suitable for the conditions. If practicable, the use of dissimilar metals in contact anywhere in the system should be avoided to eliminate the possibility of galvanic action.

### 3.5.2 Specifications

Corrosion-resistant rigid conduit is available in three general types: (1) aluminum, (2) silicon–bronze alloy, and (3) plastic coated.

Aluminum rigid conduit is useful in installations where certain chemical fumes or vapors are present that have little effect upon aluminum but have a severe corrosive effect upon steel. It is widely used because its light weight reduces installation labor costs. Uncoated or unprotected aluminum conduit should not be embedded in concrete or buried earth, particularly where soluble chlorides are present.

A silicon–bronze alloy conduit, known as Everdur electrical conduit, has special corrosion-resistant characteristics. Its use is advantageous for installations exposed to the weather as on bridges, piers, and dry docks and along the sea coast; in oil refineries and chemical plants; in sewage-disposal works, underground for water supply works; and in wiring to underwater swimming pool lighting fixtures.

Plastic-coated rigid conduit consists of standard galvanized steel conduit over the surface of which seamless coatings of polyvinyl chloride plastic have been extruded. Thus a uniform nonporous protective coating is formed over the entire length of the raceway. All couplings and fittings should be covered with plastic sleeves or carefully wrapped with three thicknesses of standard vinyl electric insulating tape. This plastic-coated conduit is flame resistant; highly resistant to the action of oils, grease, acids, alkalies, and moisture; does not oxidize, deteriorate, or shrink when exposed to sunlight and weather; and provides excellent resistance to abrasion, impact, and other mechanical wear. Typical applications are in meat-packing plants, malt-liquor industries, paper and allied industries, production plants for industrial organic and inorganic products, soap and related products industries, tanneries and leather-finishing plants, canneries, food processing industries, dairies, fertilizer plants, and petroleum industries.

The use of rigid aluminum conduit has gained wide acceptance because of its light weight; excellent grounding conductivity; ease of threading, bending, and installation; resistance to corrosion; and low losses for installed ac circuits. Installations of rigid aluminum conduit require no maintenance, painting, or protective treatment in most applications. Because of its high resistance to corrosion, this conduit should be used in many severely corrosive industrial environments such as sewerage plants, water treatment stations, filtration plants, many chemical plants, and installations around salt water.

When aluminum conduit is buried in concrete or mortar, a limited chemical reaction on the conduit surface forms a self-stopping coating. This prevents significant corrosion for the life of the structure. However, calcium chloride or similar soluble chlorides sometimes are used to speed concrete setting. In limiting the use of such speeding agents, the American Concrete Institute and most building codes recognize that embedded metals can be damaged by chlorides. As a result, if aluminum conduit is to be buried in concrete, the installer should be absolutely sure that the concrete will contain no chlorides. If there is any doubt, rigid steel conduit should be used, because, even though chlorides can damage steel conduit to some degree, this will not lead to cracking or spalling of concrete.

Although aluminum conduit can be buried safely in many soils, precautions are recommended because of unpredictable moisture, stray electric current, and chemical variations in almost every soil. To assure protection when buried directly in earth, aluminum or steel conduit should be coated with bituminous or asphalt paint, wrapped with plastic tape, or encased in a chloride-free concrete envelope.

Since it is a nonmagnetic metal, aluminum conduit reduces voltage drop in installed copper or aluminum wire up to 20 percent compared to a corresponding steel conduit installation where ac circuits are involved.

Rigid nonmetallic conduit of the heavy-wall, polyvinyl chloride (PVC) type may be used for exposed wiring in the following:

1. Portions of dairies, laundries, canneries, or other wet locations.
2. Locations where walls are frequently washed.
3. Locations subject to severe corrosive influences.
4. Damp or dry locations.
5. Locations subject to chemicals for which the PVC material is specifically approved. It may also be embedded in concrete walls, floors, and ceilings.

Liquid-tight flexible metal conduit is similar to regular flexible metal conduit except that it is covered with a liquid-tight plastic sheath. It is not intended for general-purpose wiring but has definite advantages in many cases for the wiring of machines and portable equipment. Its use is restricted by the code as follows:

1. It is allowed for connections of motors or portable and stationary equipment where flexibility of connection is required.
2. Its use is not allowed (a) where subject to physical damage; (b) where in contact with rapidly moving parts; (c) under conditions such that its temperature, with or without enclosed conductors carrying current, is above 60°C (140°F); (d) in any hazardous location, except as described in Sections 501-4(B), 502-4, and 503-3 of Article 500 of the code, unless it is specifically approved for such use.

Care must be exercised in its installation to employ only suitable terminal fittings that are approved for the purpose. Liquid-tight flexible metal conduit in sizes $1^1/_2$ in. and larger must be bonded in accordance with Section 259-79 of the code, unless specifically approved for use without a separate bond.

Electrometallic tubing (EMT) is also steel but with much thinner walls than the rigid galvanized steel. EMT is fitted together with pressure fittings and is not threaded. It is used primarily indoors.

Heavy-wall PVC conduit must not be used in the following:

1. In hazardous locations, except as permitted in Section C 514-8 and 515-5 of the National Electric Code.
2. In the concealed spaces of combustible construction.
3. For the support of fixtures or other equipment.
4. Where subject to ambient temperatures exceeding those for which the conduit is approved.
5. In sunlight unless approved for the purpose. A PVC conduit that is sunlight resistant is marked accordingly. Underwriter's Laboratories list rigid nonmetallic PVC conduit for aboveground applications with two different labels. Label A is identified by a yellow background, and label B by a white background. The conduits are manufactured in sizes from $^1/_2$ to 6 in. The extra-heavy wall type of conduit is utilized for special applications that require additional mechanical protection, such as pole risers or similar uses where conduit is subject to physical damage.
6. As described in Article 347 of the National Electric Code.

Flexible conduit and liquid-tight flexible conduit are easily bent, employ pressure fittings, and, if coated with a plastic material, are waterproof. Such conduit is generally used where short radius bends are necessary, where several bends are necessary in a short distance, or where vibration exists.

### 3.5.3 Conduit Bodies or Fittings

Conduit bodies or fittings are used to adapt conduit runs to different situations and conditions. The term *fitting* refers to many different items, all associated with conduit runs and the conditions they must meet; the most common are defined as follows:

*Conduit bodies:* For installation in conduit systems to act as pull boxes, make 90° bends, provide for splices, taps, mounting outlets, tees, x-fittings, and so on

*Cast outlet or device boxes:* For use in conjunction with threaded systems to serve as pull boxes and junction boxes, accommodate wiring devices, support light fixtures, provide openings in system for taps and splices

*Cable and cord fittings:* Cable and cord terminators into and out of bulkheads, junction or pull boxes, rigid conduit systems, and so on

*Junction boxes:* A wide variety of boxes for use as pull boxes, enclosures for splices and taps, and housings for apparatus, instruments, terminal strips, relays, and the like

*Miscellaneous fittings:* Elbows, couplings, grounding devices, plugs, reducers, service entrance fittings, unions, expansion joints, conduit seals, drains, and so on

All the preceding are manufactured in a range of sizes from $^1/_2$ to 6 in. to fit all types of conduit systems. The electrical drafter should have catalogs of the various manufacturers at his or her disposal. These illustrate the many applications and combinations available. Fittings are made with or without hubs for attachment to threaded or threadless conduit, and are manufactured in the same range of materials as conduit.

## 3.6 CONDUIT SIZING

### 3.6.1 Characteristics

All the conduits described previously come in a range of sizes from $^1/_2$ to 6 in. diameter. When we refer to the diameter, we are speaking of the *internal diameter* (ID).

When we speak of sizing conduit, we are referring to a process of determining what size of conduit is necessary to carry a specified number of wires of a given AWG size. First, we must consider a basic fact that will govern our calculations: the National Electric Code states that only 40 percent of the total internal area of a conduit can be used for wire. Thus we cannot fill a piece of conduit with wire; in fact, we can only use a little less than half of it.

This rule was made for several reasons, the more important of which are the following:

1. It would be virtually impossible to pull a bundle of wire any distance whatsoever if it filled the conduit, and it would be impossible to pull it around bends.
2. If a wire fits too tightly it would be damaged in pulling it into the conduit and might possibly short out and cause a fire.
3. Sufficient area is needed for heat dissipation.

### 3.6.2 Conduit Fill Tables

There are two ways to determine the size of conduit needed for a given installation: (1) If we know the type of wire we are using, we can look in the conduit fill tables published in the National Electric Code (see also Figure 3.15 of this text) to determine exactly what size of conduit is needed; these tables take into consideration the 40 percent fill requirement of the NEC. (2) We can use a mathematical method and calculate the areas and fill.

Let's look at the table method first; a copy of the table is shown in Figure 3.15. There are only two facts we need to know to utilize this table: (1) the AWG wire needed for our service, and (2) the type of insulation or wire we are using.

For our calculation, let's assume we need three #6 wires, copper, type THWN. In the table in the left-hand column we locate the section covering type THWN. We then look under the column headed conductor, AWG or MCM, and find #6. We now move horizontally to the right along the line with #6 wire until we find a number equal to or bigger than the number of wires we are going to use. Having located this number, we move straight up to the section marked conduit size at the top of the table and read the

**Table 3A. Maximum Number of Conductors in Trade Sizes of Conduit or Tubing**
(Based on Table 1, Chapter 9)

| Conduit Trade Size (Inches) | | ½ | ¾ | 1 | 1¼ | 1½ | 2 | 2½ | 3 | 3½ | 4 | 4½ | 5 | 6 |
|---|---|---|---|---|---|---|---|---|---|---|---|---|---|---|
| **Type Letters** | **Conductor Size AWG, MCM** | | | | | | | | | | | | | |
| TW, T, RUH, | 14 | 9 | 15 | 25 | 44 | 60 | 99 | 142 | | | | | | |
| RUW, | 12 | 7 | 12 | 19 | 35 | 47 | 78 | 111 | 171 | | | | | |
| XHHW (14 thru 8) | 10 | 5 | 9 | 15 | 26 | 36 | 60 | 85 | 131 | 176 | | | | |
| | 8 | 2 | 4 | 7 | 12 | 17 | 28 | 40 | 62 | 84 | 108 | | | |
| RHW and RHH | 14 | 6 | 10 | 16 | 29 | 40 | 65 | 93 | 143 | 192 | | | | |
| (without outer | 12 | 4 | 8 | 13 | 24 | 32 | 53 | 76 | 117 | 157 | | | | |
| covering), | 10 | 4 | 6 | 11 | 19 | 26 | 43 | 61 | 95 | 127 | 163 | | | |
| THW | 8 | 1 | 3 | 5 | 10 | 13 | 22 | 32 | 49 | 66 | 85 | 106 | 133 | |
| TW, | 6 | 1 | 2 | 4 | 7 | 10 | 16 | 23 | 36 | 48 | 62 | 78 | 97 | 141 |
| T, | 4 | 1 | 1 | 3 | 5 | 7 | 12 | 17 | 27 | 36 | 47 | 58 | 73 | 106 |
| THW, | 3 | 1 | 1 | 2 | 4 | 6 | 10 | 15 | 23 | 31 | 40 | 50 | 63 | 91 |
| RUH (6 thru 2), | 2 | 1 | 1 | 2 | 4 | 5 | 9 | 13 | 20 | 27 | 34 | 43 | 54 | 78 |
| RUW (6 thru 2), | 1 | | 1 | 1 | 3 | 4 | 6 | 9 | 14 | 19 | 25 | 31 | 39 | 57 |
| FEPB (6 thru 2), | 0 | | 1 | 1 | 2 | 3 | 5 | 8 | 12 | 16 | 21 | 27 | 33 | 49 |
| RHW and | 00 | | 1 | 1 | 1 | 3 | 5 | 7 | 10 | 14 | 18 | 23 | 29 | 41 |
| RHH (with- | 000 | | 1 | 1 | 1 | 2 | 4 | 6 | 9 | 12 | 15 | 19 | 24 | 35 |
| out outer | 0000 | | | 1 | 1 | 1 | 3 | 5 | 7 | 10 | 13 | 16 | 20 | 29 |
| covering) | 250 | | | 1 | 1 | 1 | 2 | 4 | 6 | 8 | 10 | 13 | 16 | 23 |
| | 300 | | | 1 | 1 | 1 | 2 | 3 | 5 | 7 | 9 | 11 | 14 | 20 |
| | 350 | | | | 1 | 1 | 1 | 3 | 4 | 6 | 8 | 10 | 12 | 18 |
| | 400 | | | | 1 | 1 | 1 | 2 | 4 | 5 | 7 | 9 | 11 | 16 |
| | 500 | | | | 1 | 1 | 1 | 1 | 3 | 4 | 6 | 7 | 9 | 14 |
| | 600 | | | | | 1 | 1 | 1 | 3 | 4 | 5 | 6 | 7 | 11 |
| | 700 | | | | | 1 | 1 | 1 | 2 | 3 | 4 | 5 | 7 | 10 |
| | 750 | | | | | 1 | 1 | 1 | 2 | 3 | 4 | 5 | 6 | 9 |

**FIGURE 3.15.** Conduit Fill Tables (Courtesy National Fire Protection Association, National Electric Code 1981)

**Table 3B. Maximum Number of Conductors in Trade Sizes of Conduit or Tubing**
(Based on Table 1, Chapter 9)

| Type Letters | Conductor Size AWG, MCM | ½ | ¾ | 1 | 1¼ | 1½ | 2 | 2½ | 3 | 3½ | 4 | 4½ | 5 | 6 |
|---|---|---|---|---|---|---|---|---|---|---|---|---|---|---|
| | 14 | 13 | 24 | 39 | 69 | 94 | 154 | | | | | | | |
| | 12 | 10 | 18 | 29 | 51 | 70 | 114 | 164 | | | | | | |
| THWN, | 10 | 6 | 11 | 18 | 32 | 44 | 73 | 104 | 160 | | | | | |
| | 8 | 3 | 5 | 9 | 16 | 22 | 36 | 51 | 79 | 106 | 136 | | | |
| THHN, | 6 | 1 | 4 | 6 | 11 | 15 | 26 | 37 | 57 | 76 | 98 | 125 | 154 | |
| FEP (14 thru 2), | 4 | 1 | 2 | 4 | 7 | 9 | 16 | 22 | 35 | 47 | 60 | 75 | 94 | 137 |
| FEPB (14 thru 8), | 3 | 1 | 1 | 3 | 6 | 8 | 13 | 19 | 29 | 39 | 51 | 64 | 80 | 116 |
| PFA (14 thru 4/0) | 2 | 1 | 1 | 3 | 5 | 7 | 11 | 16 | 25 | 33 | 43 | 54 | 67 | 97 |
| PFAH (14 thru 4/0) | 1 | | 1 | 1 | 3 | 5 | 8 | 12 | 18 | 25 | 32 | 40 | 50 | 72 |
| Z (14 thru 4/0) | | | | | | | | | | | | | | |
| XHHW (4 thru 500MCM) | 0 | | 1 | 1 | 3 | 4 | 7 | 10 | 15 | 21 | 27 | 33 | 42 | 61 |
| | 00 | | 1 | 1 | 2 | 3 | 6 | 8 | 13 | 17 | 22 | 28 | 35 | 51 |
| | 000 | | 1 | 1 | 1 | 3 | 5 | 7 | 11 | 14 | 18 | 23 | 29 | 42 |
| | 0000 | | 1 | 1 | 1 | 2 | 4 | 6 | 9 | 12 | 15 | 19 | 24 | 35 |
| | 250 | | | 1 | 1 | 1 | 3 | 4 | 7 | 10 | 12 | 16 | 20 | 28 |
| | 300 | | | 1 | 1 | 1 | 3 | 4 | 6 | 8 | 11 | 13 | 17 | 24 |
| | 350 | | | 1 | 1 | 1 | 2 | 3 | 5 | 7 | 9 | 12 | 15 | 21 |
| | 400 | | | | 1 | 1 | 1 | 3 | 5 | 6 | 8 | 10 | 13 | 19 |
| | 500 | | | | 1 | 1 | 1 | 2 | 4 | 5 | 7 | 9 | 11 | 16 |
| | 600 | | | | 1 | 1 | 1 | 1 | 3 | 4 | 5 | 7 | 9 | 13 |
| | 700 | | | | | 1 | 1 | 1 | 3 | 4 | 5 | 6 | 8 | 11 |
| | 750 | | | | | 1 | 1 | 1 | 2 | 3 | 4 | 6 | 7 | 11 |
| XHHW | 6 | 1 | 3 | 5 | 9 | 13 | 21 | 30 | 47 | 63 | 81 | 102 | 128 | 185 |
| | 600 | | | | 1 | 1 | 1 | 1 | 3 | 4 | 5 | 7 | 9 | 13 |
| | 700 | | | | | 1 | 1 | 1 | 3 | 4 | 5 | 6 | 7 | 11 |
| | 750 | | | | | 1 | 1 | 1 | 2 | 3 | 4 | 6 | 7 | 10 |

**FIGURE 3.15.** *continued*

**SELF-CHECK QUIZ 3.3** **Cover the right side of the page and answer the questions.**

| | |
|---|---|
| 3.17 Motor branch circuit conductors are required to have an ampacity of not less than ______% of motor full-load current. | 3.17 Ans.: *125%* Ref. 3.4.3 |
| 3.18 The NEC states that only ______% of the internal area of a conduit may be used. | 3.18 Ans.: *40%* Ref. 3.6.1 |
| 3.19 RGS means __________ __________ __________ | 3.19 Ans.: *rigid galvanized steel* Ref. 3.5.1 |
| 3.20 Conduit is a __________ | 3.20 Ans.: *raceway* Ref. 3.5. |
| 3.21 PVC is the abbreviation for __________ | 3.21 Ans.: *polyvinyl chloride* Ref. 3.5.2 |
| 3.22 Conduits range in size from __________to __________in., internal diameter. | 3.22 Ans.: $^1/_2$ to *6 in.* Ref. 3.5.2 |
| 3.23 The formula for the area of a circle is __________. | 3.23 Ans.: $A = \pi r^2$ Ref. 3.6.3 |
| 3.24 ID means __________ __________ | 3.24 Ans.: *inside diameter* Ref. 3.6.1 |

size conduit required for three #6 type THWN wires. We see that three #6 type THWN wires will fit in a $^{3}/_{4}$-in. conduit.

See if you can follow the steps and verify the following (all copper):

| *No.* | *Type* | *Conduit Size* |
|---|---|---|
| 4 #2 | RHW wires | $1^{1}/_{4}$"C. |
| 24 #10 | THW wires | $1^{1}/_{2}$"C. |
| 24 #10 | THWN wires | $1^{1}/_{4}$"C. |
| 6 #250MCM | THW wires | 3"C. |
| 6 #500MCM | THWN wires | 4"C. |
| 39 #12 | THW wires | 2"C. |
| 125 #4 | THWN wires | 6"C. |

### 3.6.3 Mathematical Method

To determine the size of conduit required for a specified number of wires requires the following information:

1. The outside diameter (OD) of the wire used.
2. A table showing internal areas of different sizes of conduits. Exhibit 6 in the Appendix furnishes the necessary factual information about rigid galvanized steel conduit, and Exhibit 15 gives us the OD of the different types and sizes of wire. With this information, we can establish a step-by-step process to follow in determining conduit size.

Let's assume we have six pieces of #4 type THWN wire, and we need to know what size of rigid galvanized steel conduit is necessary to carry this wire.

**Step 1.** From Exhibit 15 in the Appendix, in the left-hand column headed size-AWG-MCM, we move down until we come to #4.

**Step 2.** From the numeral 4 in the left-hand column, we move horizontally to the right across the page until we come to the column headed Types THWN, THHN. Here we find the OD to be 0.328 in. This table also gives us the area of the wire. If it did not, we could find the area by using the formula

$$A = \pi R^2 \quad \text{or} \quad A = \pi\left(\frac{D}{2}\right)^2$$

where
$A$ = area in square inches
$\pi$ = a constant, 3.1416
$D$ = diameter of circle
$R$ = radius of circle, which is half the diameter ($D/2$)

From the table we find the area to be 0.0845 in.$^2$. Using our formula,

$$\begin{aligned} A &= 3.1416 \times \left(\frac{0.328}{2}\right)^2 \\ &= 3.1416 \times (0.164)^2 \\ &= 3.1416 \times 0.0269 \\ &= 0.0844964 \text{ in.}^2 \end{aligned}$$

which when rounded off to four decimal places will be 0.0845, or the same as the area found in the table.

| Size | OD (in.) | ID (in.) | Area (in.$^2$) | Bending radius | | Span (ft.) |
|---|---|---|---|---|---|---|
| | | | | Sheath (in.) | No sheath (in.) | |
| $\frac{1}{2}$ | 0.840 | 0.622 | 0.30 | 6 | 4 | 10 |
| $\frac{3}{4}$ | 1.050 | 0.824 | 0.53 | 8 | 5 | 10 |
| 1 | 1.315 | 1.049 | 0.86 | 11 | 6 | 12 |
| $1\frac{1}{4}$ | 1.660 | 1.380 | 1.50 | 14 | 8 | 14 |
| $1\frac{1}{2}$ | 1.900 | 1.610 | 2.04 | 16 | 18 | 14 |
| 2 | 2.375 | 2.067 | 3.36 | 21 | 12 | 16 |
| $2\frac{1}{2}$ | 2.875 | 2.469 | 4.79 | 25 | 15 | 16 |
| 3 | 3.500 | 3.068 | 7.38 | 31 | 18 | 20 |
| $3\frac{1}{2}$ | 4.000 | 3.548 | 9.90 | 36 | 21 | 20 |
| 4 | 4.500 | 4.026 | 12.72 | 40 | 24 | 20 |
| 5 | 5.563 | 5.047 | 20.00 | 50 | 30 | 20 |
| 6 | 6.625 | 6.065 | 28.89 | 61 | 36 | 20 |

**FIGURE 3.16.** Conduit Fact Table, Rigid Galvanized Steel

Now we know the area of one piece of #4 type THWN wire, but we have six in our bundle. So we multiply by six to get the total area of our wire:

$$6 \times 0.0845 = 0.507 \text{ in.}^2$$

**Step 3.** Recall that our wire can only fill 40 percent of the area of the conduit we use, in accordance with the National Electric Code. We can determine the total area required in the conduit by multiplying the area of our wire bundle by 2.5. Since

$$40\% \times 2.5 = 100\%$$

the area of our wire bundle is

$$0.507 \times 2.5 = 1.2675 \text{ in.}^2$$

which is the area of the piece of conduit we need. Using Figure 3.16, we find the column headed area (square inches), and move straight down until we come to an area equal to or larger than the area of the wire bundle. We find this to be 1.50 in.$^2$ Move directly to the left to the column headed Trade Size and we find that we will require $1^1/_4$-in. conduit for six #4 THWN wires. This can be verified using Figure 3-15, Table 3A.

# 4 ONE-LINE DIAGRAMS

**4.1 GENERAL** In this and the following chapters, five types of electrical diagrams will be defined and discussed:

1. One-line diagrams
2. Elementary or schematic diagrams
3. Connection diagrams
4. Interconnection diagrams
5. Field wiring diagrams

In this chapter we will explore one-line diagrams, their purpose and use, the principles involved and methods of preparation, along with the symbols used and their meaning.

### 4.1.1 Purpose

One of the initial responsibilities of the electrical engineering group is to gather as much information as possible relative to the power distribution system to be designed for the project. The one-line diagram is the best available tool for recording this information in such a form that it can be utilized by the engineer and the manufacturer in determining the nature and extent of the switchgear and/or distribution equipment required.

A one-line diagram sketch is made in the preliminary discussions of any project involving switchgear. This sketch records such vital information as the following:

1. Voltage, frequency, ground connections, and short circuit potential of all power sources.

2. Desired course of the main power circuits.
3. Ratings of all generators, motors, transformers, and other apparatus in these circuits.
4. Length, size, and type of transmission lines, cables, and bus ducts.
5. Type and rating of loads connected to feeder circuits.
6. Metering equipment required.
7. Protective relay system required.

The one-line diagram sketch is often supplemented by schematic diagram sketches of the more important parts of the control equipment. With this information, specifications for the required switchgear can be formulated in enough detail to form the basis of a request for bid to the manufacturers.

### 4.1.2 Definition

We can define a one-line diagram as follows:

> A one-line diagram indicates, by means of single lines and simplified symbols, the course and component devices or parts of an electrical circuit or system of circuits.

From the definition it is apparent that the one-line diagram is primarily useful in showing the overall relations between the component elements of circuits and between the circuits themselves. One-line diagrams can be used to show this relationship in almost any circuit. However, practical experience has shown that the more complicated circuits of control apparatus, being more in the nature of circuit networks than definite channels for the transfer of energy, can be much more clearly and accurately indicated by elementary diagrams. The scope of the one-line diagram, accordingly, has been limited to main power, excitation, metering, and protective relay circuits.

One-line diagrams are prepared as finished drawings whenever the equipment is so complex as to require more than one elementary diagram, or where the overall function and performance of several units of switchgear equipment can be more clearly understood from a one-line diagram than from the elementary diagram.

## 4.2 PREPARATION OF ONE-LINE DIAGRAMS

### 4.2.1 General

To all who contribute to the preparation of one-line diagrams, this admonition is given: *Tell what you know on the one-line sketch but nothing that you do not know. Other engineers are likely to mistake your guesses for knowledge, often with disastrous results.*

The following items, if given special attention during preparation, ensure complete, accurate, and lucid diagrams.

### 4.2.2 Maintain Relative Geographic Relations

Insofar as it is feasible, within a reasonable space and without too severely congesting portions of the diagram, the approximate relative positions of stations and of main apparatus within stations should be maintained. Scaled or mapped relations cannot, of course, be maintained except in rare cases. The general pattern of such relations usually should be maintained. While a geography of the system is not essential to the preparation of a one-line diagram, especially in its preliminary stages, such knowledge does possess very definite value as an aid to the preparation of a complete diagram that will convey a maximum of useful information.

Before a diagram is put in its final form, therefore, every reasonable effort should be made to obtain, from the most authentic available source, a clear idea of the general geography of the system and the physical arrangement of major apparatus within stations. Particular attention should be paid to showing the correct order of units in switchgear assemblies, and space should be allowed for reasonable future extensions, especially where the probable extent of such extensions is known.

### 4.2.3 Avoid Duplication

The one-line diagram is a type of "diagram shorthand," and for this reason every line, symbol, figure, and letter has a definite meaning and is made to serve some definite purpose in conveying significant information. Therefore, duplication should be carefully avoided.

For example, when giving the rating of a current transformer, the abbreviation C.T. should not be used, because the symbol itself conveys this information. It is sufficient to state merely the type and rating, such as JS-1 800/5. Even the abbreviation amp. after the rating is unnecessary, because current transformers obviously can be rated only in amperes.

### 4.2.4 Use Standard Symbols

The use of standard symbols and conventions is desirable. If special features occur that cannot be accurately covered by standard symbols and conventions, care should be exercised to make entirely clear the meaning of any nonstandard symbol or convention devised to cover such features.

The symbols most frequently used on one-line diagrams are shown. They are based on American Standards for Graphical Symbols for Power Control and Measurement, Standard no. Z32.3-1946.

### 4.2.5 Show All Known Facts

Details of circuits and devices known to the author of a diagram may seem to be unimportant or even irrelevant at the time. To someone else, however, or even to the author at a later date, these details may assume major importance, and their omission may be the cause of much extra work and loss of time, or even errors and misunderstandings. No detail within the scope of the diagram, therefore, should be considered unimportant, and the rule, "When in doubt, show it," should be followed.

By checking the following list before releasing a diagram, the omission of some of the more important details can be avoided.

1. Manufacturer's type designations and ratings of devices.
2. Ratios of current and potential transformers.
3. Taps to be used on multiratio current transformers.
4. Connection of double-ratio current transformers.
5. Connections of power-transformer windings (ID, wye, delta, etc.).
6. Circuit-breaker ratings in volts and amperes and (if not indicated by the manufacturer's type designation) the interrupting rating.
7. Switch and fuse ratings in volts and amperes.
8. Functions of relays.
9. Ratings of machines and power transformers.
10. Type and number of trip coils on circuit breakers.

11. Size and type of cables.
12. Voltage, phase, and frequency of all incoming circuits; a statement should accompany this information stating whether or not the neutral of any apparatus connected to the source is grounded. If grounded, the statement should specify whether the ground is solid or through an impedance. If the latter, the value of the impedance should be given.

### 4.2.6 Show Future Plans

Frequently, special features are devised that have no purpose in connection with the system being drawn at the time, but which are included in equipment in order that it may fit into some plan for future changes or extensions. Such future plans should be set forth on the diagram either in diagrammatic form or by explanatory notes. For example, an outdoor station structure is often built with two disconnecting switches for a circuit breaker to be installed later. If shown without comment or an indication of the future breaker, the purpose of the second switch is entirely obscure.

### 4.2.7 Include Correct Title Data

Care should be exercised in the assignment of titles to one-line diagrams to the end that they accurately identify the installation. Titles should contain the following information:

1. Name of the station or substation, for example, Main Distribution Substation.
2. Name of the ultimate user, for example, Smith and Brown Mfg. Co. (the Jones Contracting Co., which might be the name of the contractors or consulting engineers purchasing the equipment, would not be the name of the ultimate user).

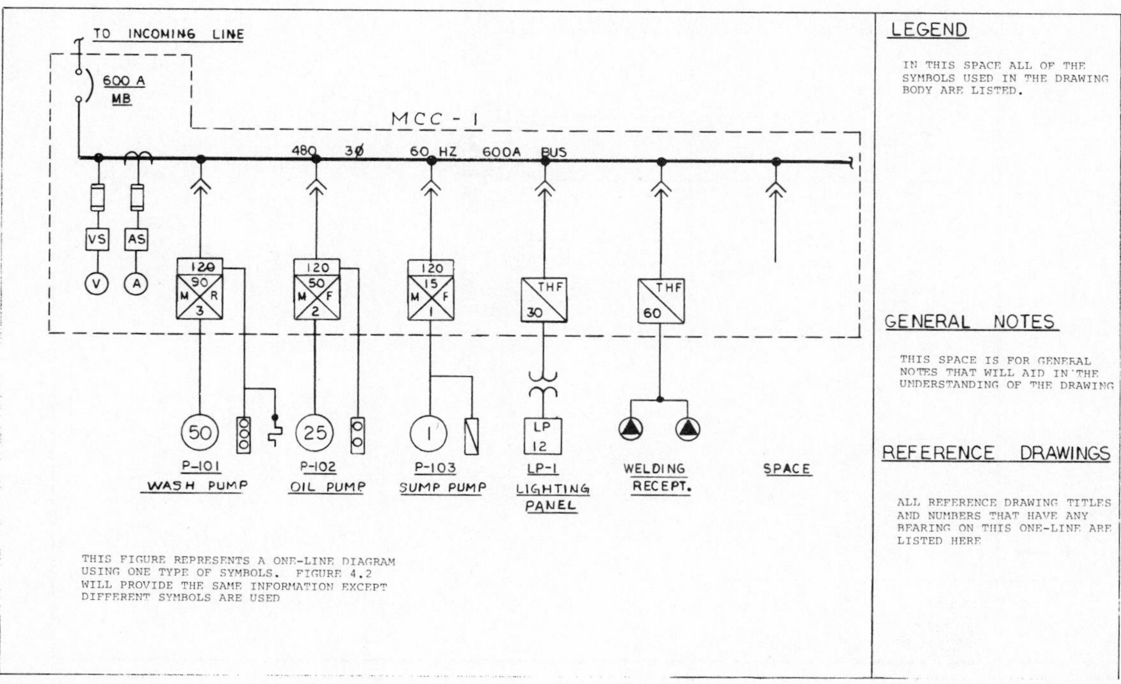


**FIGURE 4.1.** Typical One-Line Diagram

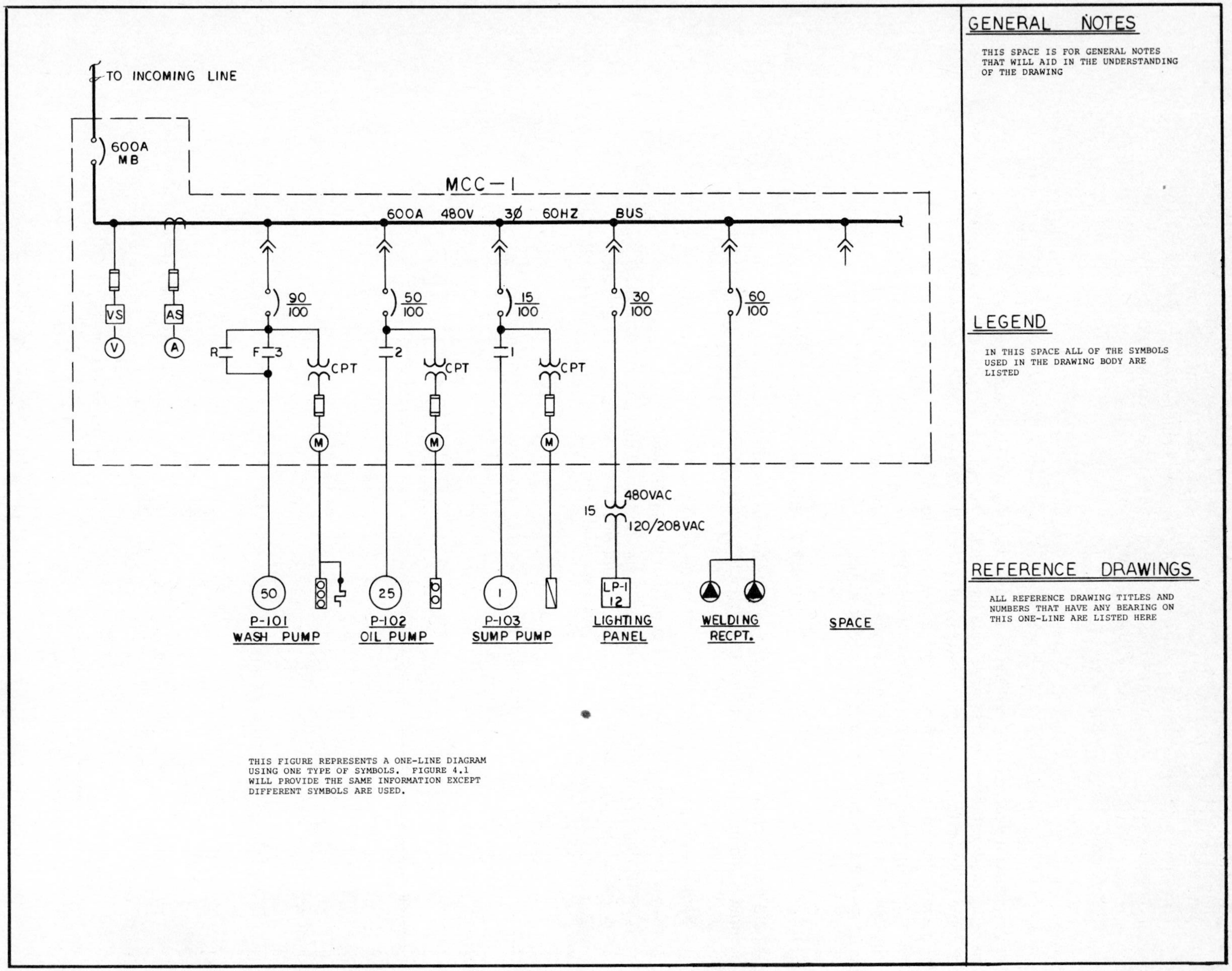

**FIGURE 4.2.** Typical One-Line Diagram

3. Name of the particular plant, for example, South Side Works.
4. Geographical location of the installation, for example, Denver, Colorado (not New York, N.Y., when that is the address of the customer's general office or purchasing agent).

## 4.3 SYMBOLS

A good knowledge of the symbols and conventions used in one-line diagrams is particularly essential to clear understanding. This form of diagram is a type of shorthand and makes use of abbreviated symbols and conventions to convey many ideas. Figures 4.1 and 4.2 are typical one-line diagrams illustrating the use of many of these symbols and conventions. The succeeding pages set forth these symbols and conventions in detail. Some of these symbols are also used in schematic wiring diagrams.

| One-Line Diagram Symbols | | | |
|---|---|---|---|
| Symbol | Description | Symbol | Description |
| R | Indicating light<br>A = Amber B = Blue<br>C = Clear G = Green<br>O = Orange P = Purple<br>R = Red Y = Yellow<br>W = White | | Fused cutout |
| A | Instrument<br>A = Ammeter<br>V = Voltmeter<br>W = Wattmeter<br>KWH = Kilowatt-hour meter | HOA | Selector switch<br>HOA = Hand-off-auto<br>LOC = Local-off-remote |
| | Protective gap | AS | Instrument switch<br>AS = Ammeter switch<br>VS = Voltmeter switch<br>TS = Test switch<br>RCS = Remote control switch |
| − + | Battery | FS | Control device<br>FS = Float switch<br>LS = Limit switch<br>LC = Level control<br>TS = Temperature switch<br>PS = Pressure switch |
| | Welding receptacle | 3<br>ST, SP, J | Pushbutton station<br>3 = No. of stations<br>OP = Open CL = Close<br>J = Jog ST = Start<br>SP = Stop O = OPF<br>SL = Slow FA = Fast |
| | Surge or lightning arrester | | High or low voltage disconnect switch, air break |

| One-Line Diagram Symbols | | | |
|---|---|---|---|
| Symbol | Description | Symbol | Description |
| | Start-stop control station | | Magnetic contractor low voltage |
| T H M S | Motor Auxiliaries<br>RTD = resistance temp. detector<br>THMS = thermistor<br>VIB = vibration switch | H O A | Hand-off-auto control station |
| VR | Voltage regulator | | High-voltage primary fuse cutout dry |
| K | Interlock<br>K = key<br>R = relay | | Manual starter low voltage |
| | Auxiliary contact normally open | | High-voltage oil-immersed disconnect switch |
| | Auxiliary contact normally closed | E | Automatic transfer switch |

| One-Line Diagram Symbols | | | |
|---|---|---|---|
| Symbol | Description | Symbol | Description |
| # | Transfer switch<br># = circuit rating | | Resistor |
| TD | Time-delay relay | | Space heater |
| | Static or surge capacitor | RH | Rheostat, variable resistor |
| | Fuse | POT | Potentiometer, variable resistor |
| | High-voltage primary fuse cutout oil | | Drawout carriage mounted fuse |
| | Semiconductor, diode rectifier | LP<br>* | Lighting panel<br>* = no. of circuits |

| One-Line Diagram Symbols | | | |
|---|---|---|---|
| Symbol | Description | Symbol | Description |
| P | Pilot device | | Connection of conductors |
| X – Y – Z | Conduit identification<br>X = P, power<br>L, lighting<br>C, control<br>Y = conduit number<br>Z = conduit size | | Cable termination |
| ALT | Alternator | | Magnetic overload |
| | Stab or separable connector | | Thermal overload |
| | Ground | | Circuit breaker |
| | Crossing or nonconnecting conductors | AF / AT | Switchgear circuit breaker, low voltage<br>AF = frame rating<br>AT = trip rating |

| One-Line Diagram Symbols | | | |
|---|---|---|---|
| Symbol | Description | Symbol | Description |
| AC / IC | Air frame circuit breaker<br>AC = amp capacity<br>IC = interrupting capacity | $V_P$ / $V_S$ | Power transformer<br>$V_P$ = primary voltage<br>$V_S$ = secondary voltage |
| X * # | Air circuit breaker<br>* = T, thermal mag.<br>M, mag. only<br>X = frame size<br># = trip element | 10 2S | 3Ø induction motor<br>10 horsepower<br>2S = 2 speed winding<br>2S2W = 2 speed, 2 windings |
| E | Electrical operator for circuit breaker | 10 | 3Ø induction motor 10 hp, vertical |
| * | Bar type current transformer<br>* = number required | | DC motor shunt wound, 1 hp |
| * | Potential transformer<br>* = number required | | DC motor series wound, 10 hp |
| | Window-type current transformer | | Wye connection with solid neutral |

| One-Line Diagram Symbols | | | |
|---|---|---|---|
| Symbol | Description | Symbol | Description |
| * | Control power transformer with drawout current limiting fuse<br>$V_P$ = primary voltage<br>$V_S$ = secondary voltage | $V_P$ kva $V_S$ CPT | Control power transformer<br>$V_P$ = primary voltage<br>$V_S$ = secondary voltage<br>kva = kilovolt-amp rating |
| CV # * O + | Air break magnetic motor starters<br>CV = control voltage<br># = trip rating<br>+ = starter size<br>* = type of CB or switch<br>O = starter type<br>F = full voltage<br>R = reversing | $V_P$ kva $V_S$ | Control power transformer with drawout current-limiting fuse |
| CV # * O + | Oil-immersed magnetic motor starter<br>(see preceding symbols) | * | Starter auxiliary contacts<br>* = number of contacts available excluding seal-in |

| One-Line Diagram Symbols | | | |
|---|---|---|---|
| Symbol | Description | Symbol | Description |
| * | Magnetic starter<br>* – type<br>R – reversing<br>RV – reduced voltage<br>2S – 2 speed<br>2SR – 2 speed reversing | $V_P$ $V_S$<br>$X_2$<br>$H_4$<br>$H_3$<br>$H_2$<br>$H_1$ $X_1$ | Transformer windings |
| | Combination magnetic starter and circuit breaker | | Magnetic starter, medium voltage |
| | Combination magnetic starter and fused disconnect | | |

**SELF-CHECK QUIZ 4.1** **Cover the right side of the page and answer the questions.**

| | |
|---|---|
| 4.1 A one-line diagram uses ________ lines and ________symbols. | 4.1 Ans.: *single* lines and *simplified* Ref. 4.1.2 |
| 4.2 One-line diagrams are prepared as ________drawings. | 4.2 Ans.: *finished* Ref. 4.1.2 |
| 4.3 Space should be allowed for ________extensions. | 4.3 Ans.: *future* Ref. 4.2.2 |
| 4.4 The use of ________symbols is desirable. | 4.4 Ans.: *standard* Ref. 4.2.4 |
| 4.5 This form of diagram is a type of ________. | 4.5 Ans.: *shorthand* Ref. 4.3 |
| 4.6 [battery symbol: + / −] indicates ________. | 4.6 Ans.: *Battery* Ref. Symbol Chart |
| 4.7 [circuit breaker symbol] indicates ________ ________. | 4.7 Ans.: *circuit breaker* Ref. Symbol Chart |
| 4.8 [start-stop station symbol] indicates ________-________ ________ ________. | 4.8 Ans.: *start-stop control station* Ref. Symbol Chart |

# SCHEMATIC WIRING DIAGRAMS

# 5

## 5.1 GENERAL

Wiring diagrams include all the devices in an electrical system and show the physical relationship of each device to the other. All poles, terminals, coils, and the like are shown in their proper place on each device. These diagrams are helpful in wiring up systems because wires can be traced and connections made exactly as shown on the diagram.

In following the electrical sequence of any circuit, however, the wiring diagram does not show the connections in a manner that can be easily followed. For this reason, a rearrangement of the circuit elements to form a schematic diagram is desirable.

## 5.2 DEFINITION

Schematic diagrams show the electrical relationship of power and control devices in a circuit. This is accomplished with single lines and symbols. Unlike wiring diagrams, which are concerned with the physical arrangement of components and their parts, schematic diagrams emphasize the function of each device and the sequence of operations of the parts, with no attempt being made to show such devices in their actual relative positions.

All control devices are shown between vertical lines, which represent the source of control power, and circuits are shown connected as directly as possible from one of these lines to the other. All connections are made in such a way that the functioning of the various devices can be easily traced. A schematic diagram thus gives the necessary information for easy following of the operation of the various devices in the circuit. It is a great aid in troubleshooting as it shows, in a simple way, the effect that opening and closing of different contacts will have on other devices in a circuit.

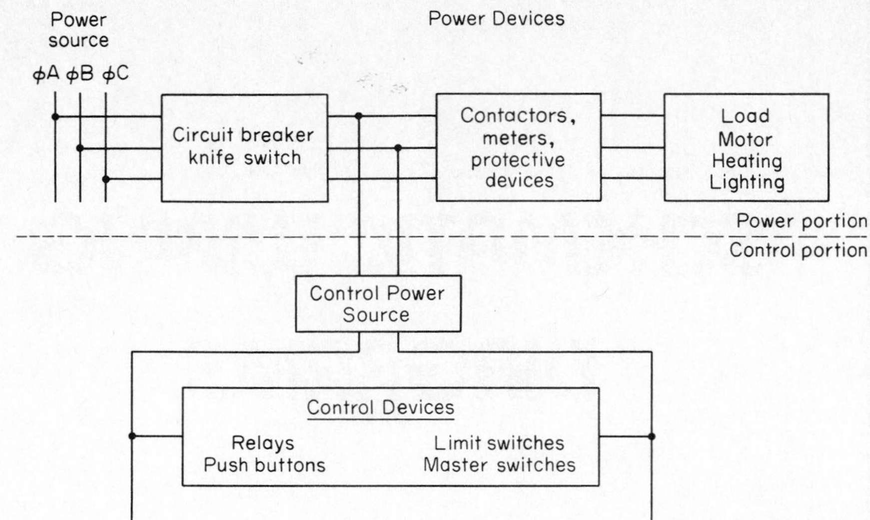


**FIGURE 5.1.** Block Schematic Diagram

## 5.3 MAIN DIVISIONS

Schematic diagrams usually consist of a power portion, drawn with heavier lines, a control power source, and a control portion. The block diagram shown in Figure 5.1 illustrates this procedure.

In some cases the power portion of the circuit and the control portion are shown on separate drawings. The power portion applies and regulates the power to an electrical load such as a motor, while the control portion controls the devices in the power circuit.

The complexity of the control and/or power distribution will dictate the space required to show a particular circuit. In addition, the control portion also includes protective devices for the machine and operator. Regardless of specific details, the orderly step-by-step approach described in the following will increase comprehension and utility of completed drawings.

## 5.4 SYMBOLS

Since the electrical circuit is represented by single lines and symbols in a schematic diagram, it is important that the student become familiar with the standard symbols used in schematic diagrams. A listing of some of the more important symbols and their meanings is part of this chapter. The majority of the symbols are standard throughout the industry and have the same meaning. Variations will appear, and unfamiliar symbols are seen as the control circuits become more complicated and the number of devices increase. All symbols are shown in their shelf or de-energized position.

## 5.5 BASIC TERMS

In addition to developing a familiarity with the basic symbols used, it will be helpful to understand the meaning of some of the terms connected with motor control. The intent of this chapter is to familiarize drafters with the terms and concepts that are fundamental to an understanding of motor control equipment and its applications. Since this book is not intended to serve as an engineering text, the material covered will be general in nature. Study of the definitions, symbols, diagrams, and illustrations, however, will give the student a sound background in the language and basic principles associated with motor control.

| Schematic Wiring Diagram Symbols | | | | |
|---|---|---|---|---|
| Component | Function | Part | Symbol | Ltr. |
| Circuit breaker | Opens or closes circuit by nonautomatic means and automatically opens circuit at a predetermined overload current | Air-insulated contacts | | CB |
| | | Oil-insulated contacts | CB | |
| Contactor | Makes and breaks power circuit to the load when coil is energized or de-energized | Coil | C | C |
| | | Normally open contact | | |
| | | Normally closed contact | | |
| Control relay | Energizes or de-energizes electrically operated devices when coil is energized or de-energized | Coil | CR | CR |
| | | Normally open contact | | |
| | | Normally closed contact | | |
| Current transformer | Induces low current in secondary wound around primary | Winding and primary | | CT |
| Current transformer with ammeter | To record induced current in secondary | Winding, primary ammeter and ground | | CT |
| Float switch | Makes or breaks when actuated by float or other liquid-level device | Float and normally open contact | | FS |
| | | Float and normally closed contact | | |

| Schematic Wiring Diagram Symbols | | | | |
|---|---|---|---|---|
| Component | Function | Part | Symbol | Ltr. |
| Foot switch | Makes or breaks contacts when manually activated | Normally open contact | | FTS |
| | | Normally closed contact | | |
| Fuse | Breaks circuit when predetermined current melts conducting element | Fuse body | * Rating | FU |
| Knife switch | Makes or breaks circuit when manually engaged or disengaged | Contacts | | KS |
| Motor starter | Makes or breaks power circuit to motor when coil is energized or de-energized | Coil overload relay and contacts | M | M |
| | | Normally closed contact | | |
| Overload relay | Disconnects motor starter at predetermined overcurrent | Magnetic coil | | OL |
| | | Thermal element | | |
| | | Contacts | | |
| Pilot light | Visually indicates presence of voltage in a circuit (color is indicated by initial) | Lamp | | PL |

| Schematic Wiring Diagram Symbols | | | | |
|---|---|---|---|---|
| Component | Function | Part | Symbol | Ltr. |
| Voltage transformer | Reduces or increases voltage when current in primary windings induces voltage in secondary windings | Windings | | XFMR |
| Ground | Connects conductor or device to ground | Grounding connection to ground | | GRND |
| Terminal | Internal terminal in motor starter to which connections are made | Terminal<br>* = terminal number | * | T |
| Solenoid valve | Electrically operated air valve, which in turn actuates a larger device | Valve and actuator | | SOL |
| Resistor | Reduces voltage in a circuit through introduction of resistance | Resistor (fixed value) | R | R |
| Rectifier | Converts alternating current to direct current | Full-wave rectifier | AC, DC +, DC −, AC | REC |

| Schematic Wiring Diagram Symbols | | | | |
|---|---|---|---|---|
| Component | Function | Part | Symbol | Ltr. |
| Push to test pilot light | Combines pilot light with self-testing circuit and switch | Lamp and switch | | |
| Pressure switch | Makes or breaks circuit when pressure or vacuum bellows actuates switch contacts | Normally open contact (NO) | | PS |
| | | Normally closed contact (NC) | | |
| Push-button switch | Makes or breaks circuit when plunger is manually depressed or released | Normally open contact (NO) | | PB |
| | | Normally closed contact (NC) | | |
| | | Combined NO and NC contacts | | |
| Rheostat | Reduces voltage in a circuit through resistance winding and tapping the reduced voltage from a sliding contact | Terminals and slider | | RH |
| Temperature switch | Makes or breaks circuit when a pre-determined temperature is attained | Normally open contact (NO) | | TS |
| | | Normally closed contact (NC) | | |
| Terminal | Provides connection point | Tie-point or splice (connection) | | |

| Schematic Wiring Diagram Symbols | | | | |
|---|---|---|---|---|
| Component | Function | Part | Symbol | Ltr. |
| Timing relay | On: Delay retards relay-contact action for predetermined time after coil is energized | Normally open timed closing (NOTC) contact | | TR OR TDR |
| | | Normally closed timed opening (NCTO) contact | | |
| | | Normally open timed open (NOTO) contact | | |
| | Off: Delay retards relay-contact action for predetermined time after coil is de-energized | Normally closed timed closed (NOTC) contact | | |
| | | No instantaneous contact | | |
| | | No instantaneous contact | | |
| Limit switch | Makes or breaks control circuit when mechanically actuated by motion or part of a powered machine | Normally open contact | | LS |
| | | Normally open contact held closed | | |
| | | Normally closed contact | | |
| | | Normally closed held open contact | | |

| Schematic Wiring Diagram Symbols | | | | |
|---|---|---|---|---|
| Component | Function | Part | Symbol | Ltr. |
| Space heater | To provide heat to cubicle or motor during periods of de-energized state to prevent moisture accumulation. | Space heater filament | | |
| Mushroom head pushbutton | To provide exaggerated surface on pushbutton actuator; used primarily as emergency switches | Head and plunger | | PB |
| Solenoid valve | Magnetic plunger, electrically actuated | Solenoid | | SOL |
| Flow switch | Makes or breaks circuit when actuated by predetermined flow or no-flow status | Normally open contact | | FLS |
| | | Normally closed contact | | |
| Hand-off automatic switch | Provides contacts to manually energize a motor or device and bypass permissive devices or to allow control devices to close contacts in auto position | Switch and contacts in off position | | HOA |
| Capacitor | (Condensor) A device consisting of two electrodes separated by a dielectric, which may be air, for introducing capacitance into a circuit | Electrodes | | CAP |

### 5.5.1 Scope

Since over 90 percent of all motors are used on AC, DC motors and their control will not be discussed. Wound-rotor motors and AC commutator motors have only a limited application and are also not included. The squirrel-cage induction motor is the most widely used motor. Therefore, its control is the subject of this chapter. The use of high voltages (2400, 4800 and higher) introduces requirements that are additional to those for 600-V equipment, and although the basic principles are unchanged, these additional requirements are not covered here.

The subject will be dealt with by first establishing guidelines for the selection of motor control equipment and then defining some basic motor control terms. The protective function of motor control is then discussed, followed by manual and magnetic control. The component parts of the standard magnetic starter are reviewed, and electrical diagrams are introduced.

### 5.5.2 Operating Characteristics of the Controller

The fundamental job of a motor controller is to start and stop the motor, and to protect the motor, machine, and operator. The controller might also be called on to provide supplementary functions, which could include reversing, jogging or inching, plugging, and operating at several speeds or at reduced levels of current and motor torque.

### 5.5.3 Environment

Controller enclosures provide protection for operating personnel by preventing accidental contact with live parts. In certain applications, the controller itself must be protected from a variety of environmental conditions such as water, rain, snow, or sleet, dirt or noncombustible dust, and/or cutting oils, coolants or lubricants. Both personnel and property require protection in environments made hazardous by the presence of explosive gases or combustible dusts.

### 5.5.4 National Codes and Standards

Motor control equipment is designed to meet the provisions of the National Electric Code (NEC). Code sections applying to industrial control devices are article 430 on motors and motor controllers and article 500 on hazardous locations.

The 1970 Occupational Safety and Health Act (OSHA) as amended in 1972, requires that each employer furnish employment free from recognized hazards likely to cause serious harm. Provisions of the act are strictly enforced by inspection.

Standards established by the National Electrical Manufacturers Association (NEMA) assist users in the proper selection of control equipment. NEMA standards provide practical information concerning construction, test, performance, and manufacture of motor control devices such as starters, relays, and contactors.

One organization that actually tests for conformity to national codes and standards is Underwriters Laboratories (UL). Equipment tested and approved by UL is listed in an annual publication, which is kept current by means of bimonthly supplements that reflect the latest additions and deletions.

### 5.5.5 Motor Controller

A motor controller includes some or all of the following functions: starting, stopping, overload protection, overcurrent protection, reversing, changing speed, jogging, plugging, sequence control, and pilot light indication. The controller can also provide the control for auxiliary equipment, such as brakes, clutches, solenoids, heaters, and signals. A motor controller may be used to control a single motor or a group of motors.

### 5.5.6 Starter

The terms "starter" and "controller" mean practically the same thing. Strictly speaking, a starter is the simplest form of controller and is capable of starting and stopping the motor and providing it with overload protection.

### 5.5.7 AC Squirrel-Cage Motor

The workhorse of industry is the ac squirrel-cage motor. Of the thousands of motors used today in general applications, the vast majority are of the squirrel-cage type. Squirrel-cage motors are simple in construction and operation; merely connect the correct power lines to the motor and it will run. The squirrel-cage motor gets its name because of its rotor construction, which resembles a squirrel cage, and has no wire winding.

### 5.5.8 Full-Load Current (FLC)

The FLC is the current required to produce full-load torque at rated speed.

### 5.5.9 Locked-Rotor Current (LRC)

During the acceleration period at the moment a motor is started, it draws a high current called the *inrush* current. This inrush current, when the motor is connected directly to the line (so that full line voltage is applied to the motor), is called the *locked rotor* or *stalled rotor* current. The locked rotor current can be from four to ten times the motor full-load current. The vast majority of motors have an LRC of about six times FLC, and therefore this figure is generally used. The six times value is often expressed as 600 percent of FLC. The code letter on the nameplate of the motor indicates the LRC of that individual motor.

### 5.5.10 Motor Speed

The speed of a squirrel-cage motor depends on the number of poles of the motor's winding. On 60 cycles, a two-pole motor runs at about 3450 rpm, a four-pole at 1725 rpm, and a six-pole at 1150 rpm. Motor nameplates are usually marked with actual full-load speeds, but frequently motors are referred to by their *synchronous speeds,* 3600, 1800, and 1200 rpm, respectively, the difference being the *slip* of the motor.

### 5.5.11 Ambient Temperature

The temperature of the air where a piece of equipment is situated is called the *ambient temperature*. Most controllers are of the enclosed type, and the ambient temperature is the temperature of the air outside the enclosure, not inside. Similarly, if a motor is said to be in an ambient temperature of 30°C (86°F), this is the temperature of the air outside the motor, not inside. NEMA standards limit both controllers and motors to a 40°C (104°F) ambient temperature.

### 5.5.12 Temperature Rise

Current passing through the windings of a motor results in an increase in the motor temperature. The difference between the winding temperature of the motor when running and the ambient temperature is called the *temperature rise*. The temperature rise produced at full load is not harmful provided the motor ambient temperature does not exceed 40°C (104°F).

Higher temperatures caused by increased current or higher ambient temperatures produce a deteriorating effect on motor insulation and lubrication. An old "rule of thumb" states that, for each increase of 10°F above the rated temperature, motor life is cut in half.

### 5.5.13 Time (Duty) Rating

Most motors have a *continuous* duty rating permitting indefinite operation at rated load. *Intermittent* duty ratings are based on a fixed operating time (5, 15, 30, 60 min), after which the motor must be allowed to cool.

### 5.5.14 Motor Service Factor

If the motor manufacturer has given a motor a *service factor,* it means that the motor can be allowed to develop more than its rated or nameplate horsepower (hp) without causing undue deterioration of the insulation. The service factor is a margin of safety. If, for example, a 10-hp motor has a service factor of 1.15, the motor can be allowed to develop 11.5 hp. The service factor depends on the motor design, and the overload is for a temporary, short-term condition only.

### 5.5.15 Jogging (Inching)

*Jogging* describes the repeated starting and stopping of a motor at frequent intervals for short periods of time. A motor would be jogged when a piece of driven equipment has to be positioned fairly closely (e.g., when positioning the table of a horizontal boring mill during set up). If jogging is to occur more frequently than five times per minute, NEMA standards require that the starter be derated. A NEMA size 1 starter has a normal duty rating of $7^1/_2$ hp at 230 V, polyphase. On jogging applications, this same starter has a maximum rating of 3 hp.

### 5.5.16 Plugging

When a motor running in one direction is momentarily reconnected to reverse the direction, it will be brought to rest very rapidly. This is referred to as *plugging*. If a motor is plugged more than five times per minute, derating of the controller is necessary, due to the heating of the contacts. Plugging can only be used if the driven machine and its load will not be damaged by the reversal of the motor torque.

### 5.5.17 Sequence (Interlocked) Control

Many processes require a number of separate motors that must be started and stopped in a definite sequence, as in a system of conveyors. When starting up, the delivery conveyor must start first with the other conveyors starting in sequence, to avoid a pile up of material. When shutting down, the reverse sequence must be followed with time delays between the shutdowns (except for emergency stops) so that no material is left on the conveyors. This is an example of a simple sequence control. Separate starters could be used, but it is common to build a special controller that incorporates starters for each drive, timer, control relay, and so on.

## 5.6 PROTECTION OF THE MOTOR

### 5.6.1 General

Motors can be damaged or their effective life reduced when subjected to a continuous current only slightly higher than their full-load current rating, times the service factor.

*Note:* Motors are designed to handle inrush or locked rotor currents without excessive temperature rise, provided the accelerating time is not too long nor the duty cycle (see *Jogging*) too frequent.

Damage to insulation and windings of the motors can also be sustained on extremely high currents of short duration, as found in grounds and short circuits.

All currents in excess of full-load current can be classified as overcurrents. In general, however, a distinction is made based on the magnitude of the overcurrent and equipment to be protected.

An overcurrent up to locked rotor current is usually the result of a mechanical overload on the motor. The subject of protection against this type of overcurrent is covered in Article 430 (part C) of the National Electric Code. In our discussion the designation "motor running overcurrent (overload) protection" will be shortened simply to *overload protection* and will define protection against overcurrents not exceeding locked rotor current.

Overcurrents due to short circuits or grounds are much higher than locked rotor currents. Equipment used to protect against damage due to this type of overcurrent must not only protect the motor but also the branch circuit conductors and the motor controllers. Provisions for the protective equipment are specified in Article 430 under part D. In our discussion we will use the term *overcurrent protection* to designate protection against high overcurrents as would typically be encountered in short circuit or grounds.

Motor overload protection differs from overcurrent protection, and each will be separately covered in succeeding sections.

### 5.6.2 Overcurrent Protection

The function of the overcurrent protective device is to protect the motor branch circuit conductors, control apparatus, and motor from short circuits or grounds. The protective devices commonly used to sense and clear overcurrents are thermal magnetic circuit breakers and fuses. The short-circuit and ground-fault protection device must be capable of carrying the starting current of the motor, *but* the device setting must not exceed the values given in Table 430-152 of the National Electric Code, depending on the type of motor and the code letter of the motor. Where the value is not sufficient to carry the starting current, it may be increased, but it must in no case exceed 400 percent of the motor full-load current.

The National Electric Code requires (with a few exceptions) a means to disconnect the motor and controllers from the line, in addition to an overcurrent protective device to clear short-circuit groundfaults. The circuit breaker incorporates fault protection and disconnect in one basic device. When overcurrent protection is provided by fuses, a disconnect switch is required, and the switch and fuses are generally combined.

### 5.6.3 Overloads

A motor has no intelligence and will attempt to drive any load, even if excessive. Exclusive of inrush or locked rotor current when accelerating, the current drawn by the motor when running is proportional to the load, varying from a no-load current (approximately 40 percent of FLC) to the full-load current rating stamped on the motor nameplate. When the load exceeds the torque rating of the motor, it draws higher than full-load current, and the condition is described as an overload. The maximum overload exists under locked rotor conditions, in which the load is so excessive that the motor stalls or fails to start and, as a consequence, draws continual inrush (LRC) of electrical overloads.

### 5.6.4 Overload Protection

The effect of an overload is a rise in temperature in the motor windings. The larger the overload, the more quickly the temperature will increase to a point damaging to the insulation and lubrication of the motor. An inverse relationship, therefore, exists between current and time: the higher the current, the shorter the time before motor damage or burn out can occur.

All overloads shorten motor life by deteriorating the insulation. Relatively small overloads of short duration cause little damage, but if sustained, they could be just as harmful as overloads of greater magnitude.

The ideal overload protection for a motor is an element with current-sensing properties very similar to the heating curve of the motor, which would act to open the motor circuit when full-load current is exceeded. The operation of the protective device should be such that the motor is quickly removed from the line when an overload has persisted too long.

### 5.6.5 Fuses

Fuses are not designed to provide overload protection. Their basic function is to protect against short circuits (overcurrents). Motors draw a high inrush current (generally six times the normal FLC) when starting. Single-element fuses have no way of distinguishing between this temporary and harmless inrush current and a damaging overload. Thus, a fuse chosen on the basis of motor FLC would blow everytime the motor started. On the other hand, if a fuse were chosen large enough to pass the starting or inrush current, it would not protect the motor against small, harmful overloads that might occur later.

Dual-element or time-delay fuses can provide motor overload protection, but suffer the disadvantage of being nonrenewable and must be replaced.

### 5.6.6 Overload Relays

The overload relay is the heart of motor protection. Like the dual-element fuse, the overload relay has inverse trip time characteristics, permitting it to hold in during the accelerating period (when inrush current is drawn), yet providing protection against continuing small overloads above FLC when the motor is running. Unlike the fuse, the overload relay is renewable and can withstand repeated trip and reset cycles without need of replacement. It should be emphasized that the overload relay does not provide short-circuit protection. This is the function of short-circuit overcurrent protective equipment like fuses and circuit breakers.

The overload relay consists of a current-sensing unit connected in the line to the motor, plus a thermomechanism, actuated by the sensing unit, that serves to directly or indirectly break the circuit. In a manual starter, an overload trips a mechanical latch, causing the starter contacts to open and disconnect the motor from the line. In magnetic starters an overload opens a set of contacts within the overload relay itself. These contacts are wired in series with the starter coil in the control circuit of the magnetic starter. Breaking the coil circuit causes the starter contacts to open, disconnecting the motor from the line.

Overload relays can be classified as being either thermal or magnetic. Magnetic overload relays react only to current excesses and are not affected by temperature. As the name implies, thermal overload relays rely on the rising temperatures caused by the overload current to trip the overload mechanism. Thermal overload relays can be further subdivided into two types: *melting alloy* and *bimetallic*.

## 5.7 MANUAL STARTER

### 5.7.1 Definition

A *manual starter* is a motor controller whose contact mechanism is operated by a mechanical linkage from a toggle handle or push button, which is in turn operated by hand. A thermal unit and direct-acting overload mechanism provide motor running overload protection. Basically, a manual starter is an on–off switch with overload relays. Manual starters are generally used on small machine tools, fans and blowers, pumps, compressors, and conveyors. They are the lowest in cost of all motor starters, have a simple mechanism, and provide quiet operation with no ac magnet hum. Moving a handle or pushing the start button closes the contacts, which remain closed until the handle is moved to *off,* or the stop button is pushed, or the overload relay thermal units trip.

### 5.7.2 Types

Manual starters are of the *fractional hp* type or the *integral hp* type and usually provide across-the-line starting. Standard manual starters cannot provide low-voltage protection or low-voltage release. If power fails, the contacts remain closed, and the motor will restart when power returns. This is an advantage for pumps, fans, compressors, oil burners, and the like, but for other applications it can be a disadvantage and *can even be dangerous* to personnel or equipment.

### 5.7.3 Magnetic Control

A high percentage of applications requires the controller to be capable of operation from remote locations or of providing automatic operation in response to signals from pilot devices such as thermostats, pressure or float switches, and limit switches. Low-voltage release or protection might also be desired. Manual starters cannot provide this type of control, and therefore magnetic starters are used.

The operating principle that distinguishes a magnetic from a manual starter is the use of an electromagnet. The electromagnet consists of a coil of wire placed on an iron core. When current flows through the coil, the iron of the magnet becomes magnetized, attracting the iron bar, called the *armature*.

The field of the permanent magnet, however, will hold the armature against the pole faces of the magnet indefinitely, and the armature could not be dropped out except by a loss of emf or by physically pulling it away. In the electromagnet, interrupting the current flow through the coil of wire causes the armature to drop out due to the presence of an air gap in the magnetic circuit.

## 5.8 MAGNETIC STARTER

### 5.8.1 Control Circuit

The circuit to the magnet coil that causes a magnetic starter to pick up and drop out is distinct from the power circuit. Although the power circuit can be single phase or polyphase, the coil circuit is always a single-phase circuit. Elements of a coil circuit include the following:

1. Magnet coil.
2. Contact(s) of the overload relay assembly.

3. Momentary or maintained contact pilot device, such as a push-button station, or pressure, temperature, liquid level or limit switch.
4. In lieu of a pilot device, the contact(s) of a relay or timer.
5. Auxiliary contact on the starter, designated as a holding circuit interlock, which is required in certain control schemes.

The coil circuit is generally identified as the control circuit, and contacts in the control circuit handle the coil load.

### 5.8.2 Two-Wire Control

In the wiring and elementary diagrams shown in Figure 5.2, two wires connect the control device (which could be a thermostat, float switch, limit switch or other maintained contact device) to the magnetic starter. When the contacts of the control device close, they complete the control circuit of the starter coil, causing it to pick up and connect the motor to the lines. When the control device contacts open, the starter is de-energized, stopping the motor.

Two-wire control provides low-voltage release, but not low-voltage protection. Wired as illustrated, the starter will function automatically in response to the direction of the control device, without the attention of an operator.

The dashed portion shown in Figure 5.2 represents the holding circuit interlock furnished on the starter, but not used in two-wire control. For greater simplicity, this portion is omitted from the conventional two-wire elementary diagram.

### 5.8.3 Three-Wire Control

A three-wire control circuit, shown in Figure 5.2, uses momentary contact start–stop push buttons. A holding circuit interlock normally open (N.O.) contact, wired in parallel with the start button, maintains the circuit. Pressing the N.O. start button completes the circuit to the coil. The power circuit contacts in lines 1, 2, and 3 close, completing the circuit to the motor, and the holding circuit contact (mechanically linked with the power contacts) also closes. Once the starter has picked up, the start button can be released, as the now closed interlock contact provides an alternate current path around the reopened start contact.

Pressing the normally closed (N.C.) stop button will open the circuit to the coil, causing the starter to drop out. An overload condition, which causes the overload contact to open, a power failure, or a drop in voltage to less than the seal-in value, would also de-energize the starter. When the starter drops out, the interlock contact reopens, and both current paths to the coil, through the start button and the interlock, are now open.

Since three wires from the push-button station are connected into the starter, at points 1 (L1), 2, and 3, this wiring scheme is commonly referred to as three-wire control.

### 5.8.4 Holding Circuit Interlock

The holding circuit interlock is a normally open (N.O.) auxiliary contact provided on standard magnetic starters and contactors. It closes when the coil is energized to form a holding circuit for the starter after the start button has been released. As a matter of economics, vertical action contactors and starters in the smaller NEMA sizes (size 0 and 1) have a holding interlock that is physically the same size as the power contacts.

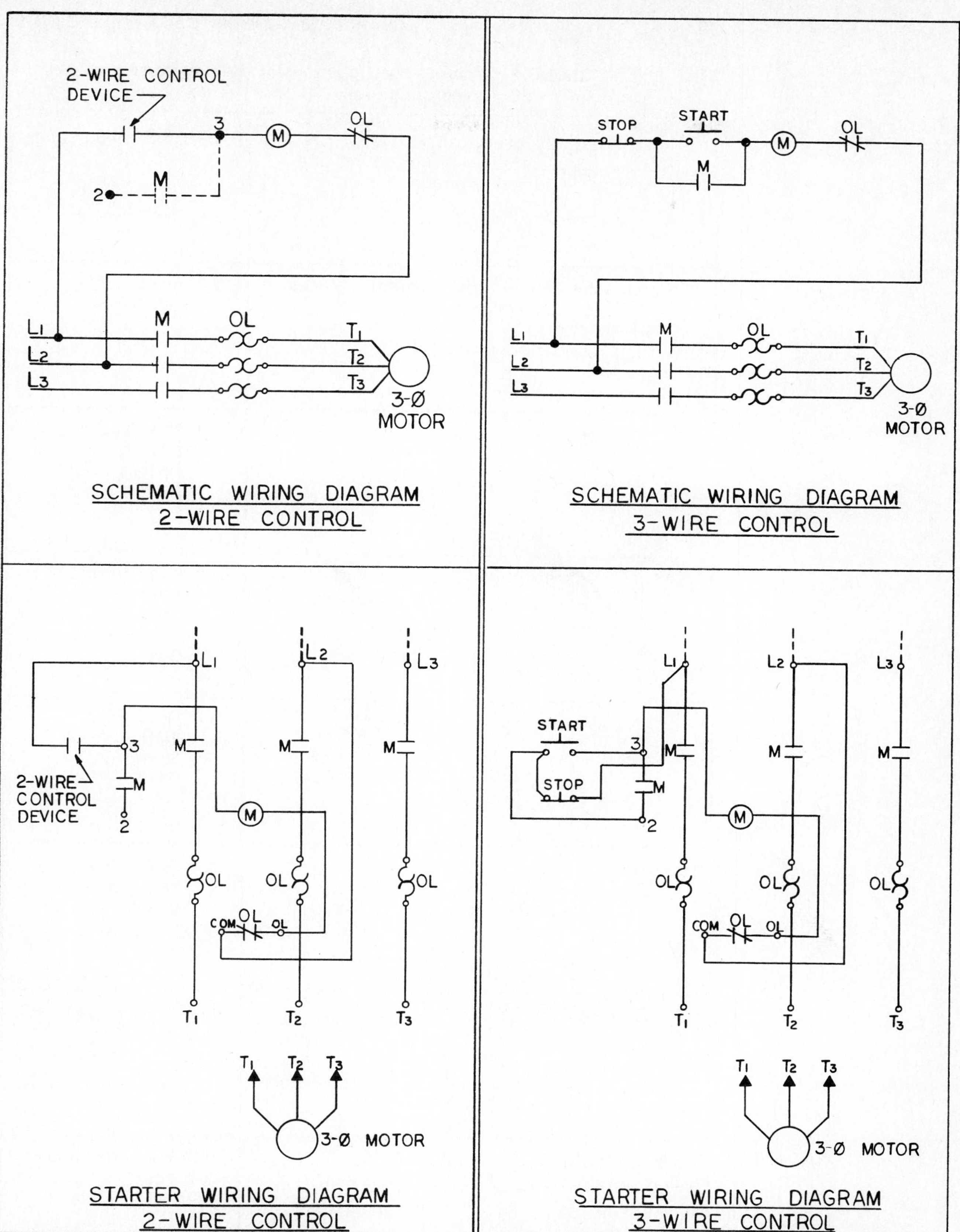


**FIGURE 5.2.** Schematic Wiring Diagrams

**SELF-CHECK QUIZ 5.1** **Cover the right side of the page and answer the questions.**

|  |  |
|---|---|
| 5.1 Schematic diagrams show the ___________relationship of power and control devices in a circuit. | 5.1 Ans.: *electrical* Ref. 5.2 |
| 5.2 In a schematic wiring diagram, devices ___________shown in are/are not their relative position. | 5.2 Ans.: *are not* Ref. 5.2 |
| 5.3 (symbol) represents a ___________ ___________. | 5.3 Ans.: *circuit breaker* Ref. Symbol Chart |
| 5.4 (symbol) represents a ___________ ___________ ___________. | 5.4 Ans.: *normally open contact* Ref. Symbol Chart |
| 5.5 (symbol) represents a ___________. | 5.5 Ans.: *fuse* Ref. Symbol Chart |
| 5.6 The job of a controller is to ___________, ___________, and ___________the motor. | 5.6 Ans.: *start, stop,* and *protect* Ref. 5.5.2 |
| 5.7 Overloads are ___________devices. | 5.7 Ans.: *protective* Ref. 5.6.3 |
| 5.8 The holding circuit interlock is a ___________ ___________ contact. | 5.8 Ans.: *normally open* Ref. 5.8.3 |

### 5.8.5 Electrical Interlocks

In addition to the main or power contacts that carry the motor current and the holding circuit interlock, a starter can be provided with externally attached auxiliary contacts, commonly called *electrical interlocks*. Interlocks are rated to carry only control circuit currents, not motor currents; N.O. and N.C. versions are available.

Among a wide variety of applications, interlocks can be used to control other magnetic devices where sequence operation is desired, to electrically prevent another controller from being energized at the same time such as forward–reverse, and to make and break circuits to indicate or alarm devices such as pilot lights, bells, or other signals. Electrical interlocks are packaged in kit form and can be easily added in the field.

### 5.8.6 Control Device (Mechanical Pilot Device)

A device that is operated by some nonelectrical means (such as the movement of a lever) and that has contacts in the control circuit of a starter is called a *control device*. Operation of the control device will control the starter and hence the motor. Typical control devices are control stations, limit switches, foot switches, pressure switches, and float switches. The control device may be of the maintained contact or momentary contact type. Some control devices have a horsepower rating and are used to directly control small motors through the operation of their contacts. When used in this way, separate overload protection (such as a manual starter) normally should be provided, as the control device does not usually incorporate overload protection.

### 5.8.7 Maintained Contact

A maintained contact control device, when operated, will cause a set of contacts to open (or close) and stay open (or closed) until a deliberate reverse operation occurs. A conventional thermostat is a typical maintained contact device. Maintained contact control devices are used with two-wire control.

### 5.8.8 Momentary Contact

A standard push button is a typical momentary contact device. Pushing the button will cause N.O. contacts to close and N.C. contacts to open. When the button is released, the contacts revert to their original positions. These types of momentary contact devices are used with three-wire control or jogging service.

### 5.8.9 Low-Voltage (Undervoltage) Release

By the nature of its control circuit connections, a two-wire control scheme provides low-voltage release. The term describes a condition in which a reduction or loss of voltage will stop the motor, but in which motor operation will automatically resume as soon as power is restored.

### 5.8.10 Low-Voltage (Undervoltage) Protection

In both two- and three-wire control, the starter will drop out and the motor will stop in response to a low-voltage condition or power failure. When power is restored, however, the starter connected for three-wire control will not pick up, as the reopened holding circuit contact and the N.O. start button contact prevent current flow to the coil. To restart the motor after a power failure, the low-voltage protection offered by three-wire control requires that the start button be depressed. A deliberate action must be performed. This ensures greater safety than that provided by two-wire control. Manual starters with low-voltage protection offer this feature.

## 5.9 FULL-VOLTAGE (ACROSS THE LINE) STARTER

As the name implies, a full-voltage or across-the-line starter directly connects the motor to the lines. The starter can be either manual or magnetic. A motor connected in this fashion draws full inrush current and develops maximum starting torque so that it accelerates the load to full speed in the shortest possible time. Across-the-line starting can be used wherever this high inrush current and starting torque are not objectionable.

With some loads the high starting torque will damage belts, gears, couplings, and materials being processed. High inrush current can produce line voltage dips that cause lamp flicker and disturbances to other loads. Lower starting currents and torques are therefore often required and are achieved by reduced-voltage starting.

## 5.10 COMMON CONTROL

The coil circuit of a magnetic starter or contactor is distinct from the power circuit. The coil circuit could be connected to any single-phase source of power, and the controller would be operable, provided the coil voltage and frequency match the service to which it is connected.

When the control circuit is tied back to lines 1 and 2 of the starter, the voltage of the control circuit is always the same as the power circuit voltage and the term *common control* is used to describe the relationship. Other variations include separate control and control through a control power transformer, stepping down the power circuit voltage to a safer, lower voltage.

## 5.11 CONTACTOR

### 5.11.1 Definition

The general classification of contactor covers a type of electromagnetic apparatus designed to handle relatively high currents. A special form of contactor exists for lighting load applications.

The conventional contactor is identical in appearance, construction, and current-carrying ability to the equivalent NEMA-sized magnetic starter. The magnet assembly and coil, contacts, holding circuit interlock, and other structural features are the same. The significant difference is that the contactor does not provide overload protection. Contactors, therefore, are used in motor circuits if overload protection is separately provided. A typical application of the latter is in a reversing starter.

### 5.11.2 Lighting Contactor

Filament-type lamps (tungsten, infrared, quartz) have inrush currents approximately 15 to 17 times the normal operating currents. Standard motor control contactors must be derated if used to control this type of load, to prevent welding of the contacts on the high initial current.

A NEMA size 1 contactor has a continuous current rating of 27 a, but if used to switch certain lighting loads, it must be derated to 15 a. The standard contactor, however, need not be derated for resistance heating or fluorescent lamp loads, which do not impose as high an inrush current.

### 5.11.3 Standard Compared to Lighting Contactors

Lighting contactors differ from standard contactors in that the contact tip material is a silver tungsten carbide that resists welding on high initial currents. A holding circuit interlock is not normally provided, since this type of contactor is frequently controlled by a two-wire pilot device such as a time clock or photoelectric relay.

Unlike standard contactors, lighting contactors are not horsepower rated or categorized by NEMA size, but are designated by ampere ratings (20, 30, 60, 100, 200, and 300 a). It should be noted that lighting contactors are specialized in their application and should not be used on motor loads.

## 5.12 MECHANICALLY HELD CONTACTS

In a conventional contactor, current flow through the coil creates a magnetic pull to seal in the armature and maintain the contacts in a switched position (N.O. contacts will be held closed; N.C. will be held open). Because the contactor action is dependent on the current flow through the coil, the contactor is described as electrically held. As soon as the coil is de-energized, the contacts will revert to their initial position.

Mechanically held versions of contactors are also available. The action is accomplished through use of two coils and a latching mechanism. Energizing one coil (latch coil) through a momentary signal causes the contacts to switch, and a mechanical latch holds the contacts in this position, even though the initiating signal is removed, and the coil is de-energized. To restore the contacts to their initial position, a second coil (unlatch coil) is momentarily energized.

Mechanically held contactors and relays are used where the slight hum of an electrically held device would be objectionable, as in auditoriums, hospitals, and churches.

## 5.13 COMBINATION STARTER

A combination starter is so named since it combines a disconnect means, which might incorporate a short circuit ground-fault protective device, and a magnetic starter in one enclosure. Compared with a separately mounted disconnect and starter, the combination starter takes up less space, requires less time to install and wire, and provides greater safety. Mechanical interlocks are provided on all combination starters. Safety to personnel is assured because the door is mechanically interlocked, so it cannot be opened without first opening the disconnect.

Combination starters can be furnished with circuit breakers or fuses to provide overcurrent protection, and are available in nonreversing and reversing versions. They may also contain the control power transformer as part of the starter.

## 5.14 SEQUENCE OF OPERATION

With the preceding definitions in mind, and having established a familiarity with the basic symbols used, let's examine a typical motor schematic diagram with basic power distribution and start–stop control, observing the symbols used and following the sequence of operations.

Power control devices handle high current and are generally few in number on a control panel. However, location in the schematic is important and will determine the wiring layout and arrangement of the panel. Consequently, the power portion is drawn first. Figure 5.3 represents a typical motor schematic diagram with basic power distribution and start–stop control.

Beginning with the power portion, the three vertical lines ① on the left-hand side represent the power bus or source of power for our motor. The three horizontal lines labeled L1, L2, and L3 ② represent an extension of that power source to the power

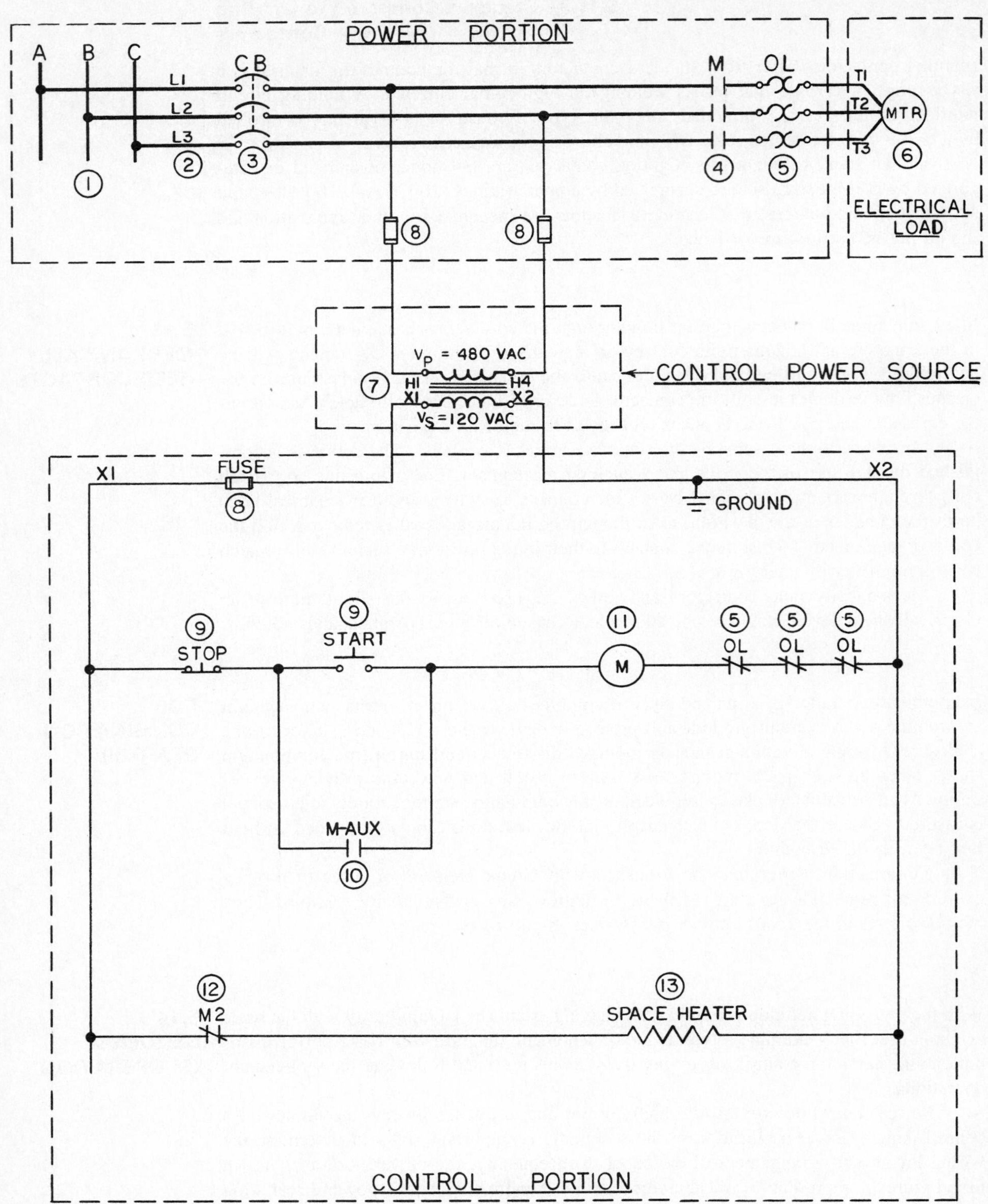


**FIGURE 5.3.** Schematic or Elementary Diagram

**SELF-CHECK QUIZ 5.2 Cover the right side of the page and answer the questions.**

| | |
|---|---|
| 5.9 A full-voltage starter is sometimes called an ________ ________ ________starter. | 5.9 Ans.: *across the line* Ref. 5.8 |
| 5.10 The three parts of a schematic diagram are ________, ________ and ________. | 5.10 Ans.: *power, control power,* and *control* devices Ref. 5.3 |
| 5.11 A combination starter combines a means of ________and ________. | 5.11 Ans.: *Disconnect protection* and *starter* Ref. 5.13 |
| 5.12 All devices in a schematic are shown in their ________or ________ position. | 5.12 Ans.: *shelf* or *de-energized* Ref. 5.4 |
| 5.13 [symbol] represents a ________. | 5.13 Ans.: *float switch* Ref. Symbol Chart |
| 5.14 [symbol: *] represents a ________. | 5.14 Ans.: *terminal* Ref. Symbol Chart |
| 5.15 Controller enclosures provide protection for ________ ________. | 5.15 Ans.: *operating personnel* Ref. 5.5.3 |
| 5.16 Motor control equipment is designed to meet the provisions of the ________. | 5.16 Ans.: NEC Ref. 5.5.4 |

portion of our circuit. A circuit breaker ③ serves as a means of disconnect as well as protection for the circuit. The circuit breaker will completely isolate the power and control portions from the source when in open position.

The three power contacts ④ close when the coil of the contactor ⑪ is energized, connecting the load ⑥, in this case a motor, directly with the source. The overloads ⑤, which are thermally motivated devices dependent on set temperature ranges, are closed at all times. In the event of motor heat-up from overloading, drag, or any of a number of factors, these devices will open when the temperature reaches a preset limit, opening the coil circuit and isolating the load from the power source through the M contacts. These are specifically equipment protective devices.

Examining the control power source, the small transformer ⑦ is the control power transformer, CPT, and reduces the line voltage, in this case 480 Vac to 120 Vac for our control voltage. Control voltages can be any voltage up to 600 Vac but are generally 120 Vac for safety reasons. Fuses ⑧ protect the transformer and control circuit. In the event of grounding or short circuit, these fuses will open or "blow," isolating the control circuit.

From the secondary side of the CPT, two wires, X1 and X2, form the hot and neutral buses for our control devices. The start–stop station ⑨ is one device. The stop portion is a normally closed, momentary contact device, and the start portion is a normally open momentary contact device. See Section 5.8.8. Since the start contact is momentary, some means of maintaining a current path is necessary. This function is accomplished with the M-AUX contact ⑩. When the start button is depressed, the coil M is energized, and the M-AUX ⑩ will close and remain closed as long as the coil is energized. When external pressure is released from the start button, it returns to its normally open position. The current path through this device is now open, and *without* the M-AUX contact ⑩ the motor would stop. However, since the coil was energized and M-AUX was closed, the current path is now through M-AUX and the motor continues to run even though the start contact is in open position. See Section 5.8.4. To stop the motor, the stop button is depressed, opening the circuit to coil M. When coil M is de-energized, contact M-AUX opens and the current path is broken. To start the motor again, merely depress the start button and the process is repeated.

In larger motors, a space heater ⑬ is provided to keep moisture out of the windings

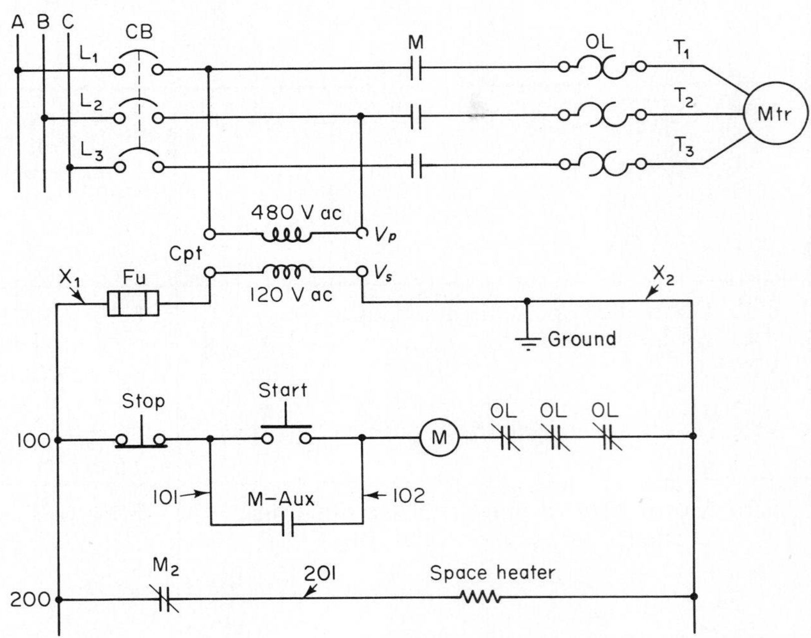


**FIGURE 5.4.** Assigning Wire Numbers to the Schematic

when the motor is *not* running. The motor, when running, will generate enough heat to keep it dry; therefore no additional heat is needed then.

Auxiliary contact M2 ⑫, a normally closed contact, provides a current path to the space heater. When coil M ⑪ is energized, contact M2 opens and the space heater is off. When the motor is stopped, coil M is de-energized, contact M2 closes, and the space heater once again comes on as long as the motor is stopped.

This is the basis of simple motor control. Additional devices, lights, alarms, and

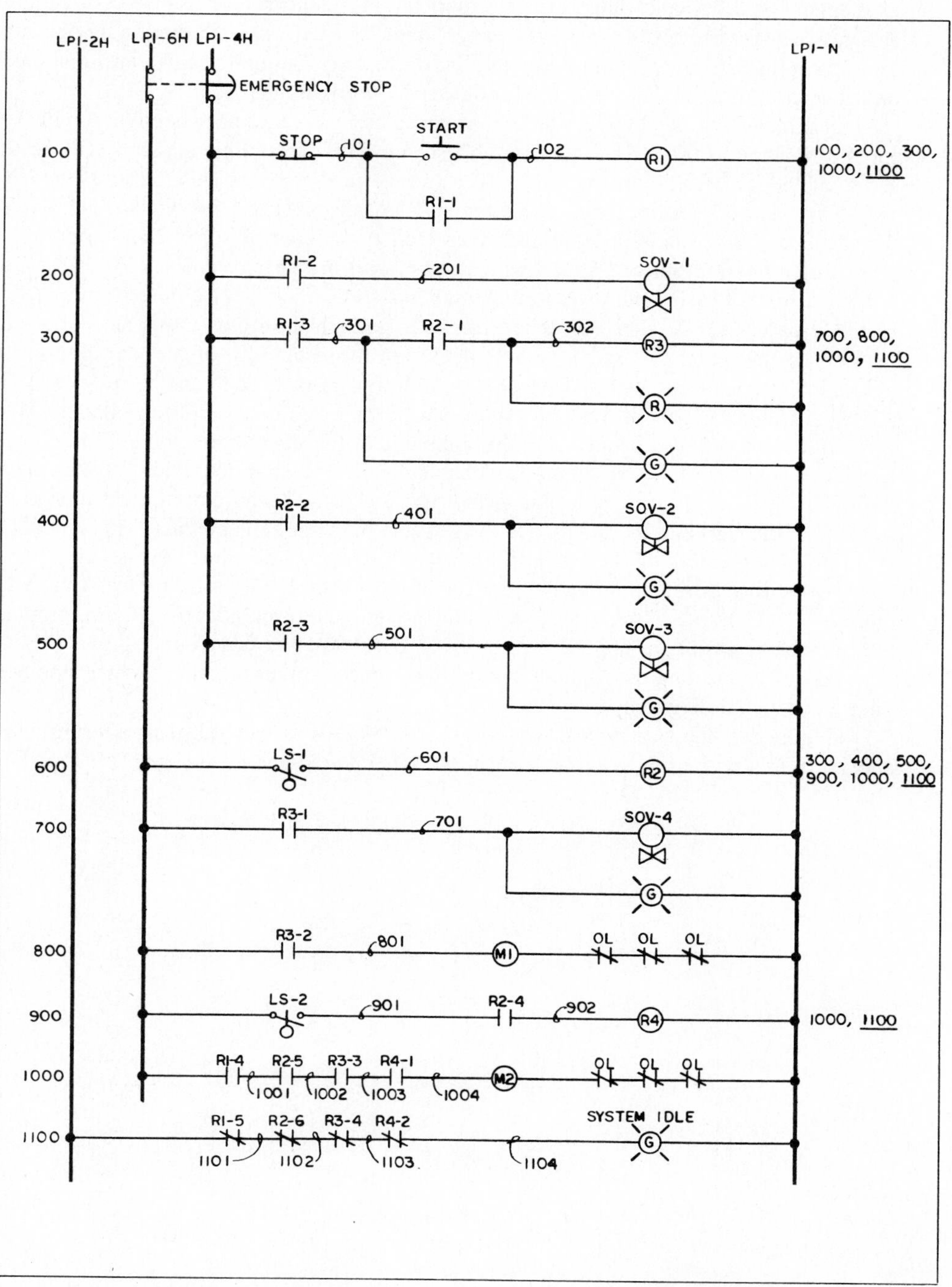


**FIGURE 5.5.** Wire Numbers on a Ladder Diagram

the like, can be added. They also will be activated by contacts or relays related to the energized or de-energized state of coil M.

## 5.15 WIRE NUMBERS

Without some means of identifying each wire in a group, whether that group consists of 2 or 200 wires, it would be extremely difficult and time consuming to assure that each wire was correctly terminated. To overcome this problem, wires are given numbers for identification. Each wire is assigned a number that is unique for that wire only. In the elementary or schematic diagram, a method of line identification is used for number assignment. A wire number will not change *unless* the wire goes through a device such as a relay coil, a contact, start or stop button, or the like. Going through a terminal block or connection to another wire will not change the wire number.

In Figure 5.4, each horizontal line has been assigned a number beginning with 100 for the first line, 200 for the second, and so on. This three-digit numeral will be the series of the individual wire numbers. The two vertical wires or bus are numbered X1 on the left and X2 on the right. These are standard designations for the two leads from the secondary of a single-phase transformer such as the control power transformer.

Remembering our rule that a wire number will *not* change unless it goes through a device, we can assign numbers to the wires in the schematic. X1 comes from one side of the transformer. At line 100, a tap from X1 goes to the right to one side of a stop button. To this point the wire number is X1. On the other side of the stop button the wire number changes to 101 and remains such until it goes through another device. Wire 101 runs from one side of the stop button and to one side of the hold-in contact. Wire 102 is the next in line and runs from the right side of the start button and hold-in contact to the coil in the starter. These numbers enable the electrician in the field to make correct terminations. He will hook wire 101 to the right side of the stop button and run it to the left side of the start button, and then to the left side of the holding contact. This procedure is followed in numbering the wires in the second line, the third line, and so on.

As further differentiation, the number of the motor is used as a prefix to the wire number. For example, if this was a schematic of pump motor number p-100, the complete wire number would be p-100-X2 or p-100-X1 or p-100-101, and so on.

Wires connecting devices that are prewired in the starter are not assigned numbers on the schematic diagram.

Figure 5.5 shows a typical ladder diagram. The same method of numbering wires applies to this type of diagram as used in the schematic diagram in Figure 5.4.

# WIRING AND CONNECTION DIAGRAMS

## 6.1 GENERAL

From the previous section we saw that a schematic diagram, sometimes referred to as an elementary diagram, is a diagram showing the devices of the circuit and the path or flow of current and is represented by single lines and symbols. It was also pointed out that the schematic diagram depicted the circuit in its simplest form and disregarded the physical placement or relationship of the devices within the circuit.

We also investigated the one-line diagram, which is similar to the schematic diagram and is defined as a diagram that indicates, by means of single lines and simplified symbols, the course and component devices of an electrical circuit or system of circuits. The schematic diagram usually depicts control circuits, while the one-line diagram is generally concerned with power and metering circuits.

We will now take a look at another type of diagram, the wiring diagram.

## 6.2 COMPARISONS

We can make the following comparisons between wiring diagrams and the previously discussed schematic diagrams.

1. *Schematic diagrams:* Use symbols to represent the various devices and components of a circuit.
   *Wiring diagrams:* Devices are shaped similar to the device itself, and shows terminals, terminal numbers, and connections to these terminals; basically, a pictorial presentation.
2. *Schematic diagrams:* Use one line to represent the circuit path regardless of the number of wires actually used or needed.
   *Wiring diagrams:* Show a line for every wire in the circuit, where it originates and where it terminates, and its path from point to point.

3. *Schematic diagrams:* Have no concern with the physical appearance, location, or relationship of devices within the circuit, but are very useful in troubleshooting or for quick reference to the circuit operation.
   *Wiring diagrams:* Attempt to show, as much as possible, the devices as they physically relate to each other, their position in the circuit and, also, to a certain degree how they actually look complete with terminals, coils, wires, and so on.

Wiring diagrams are extremely helpful to the individual actually doing the wiring of a panel, device, box, or piece of equipment, for each wire can be traced to its connection point, each device is identified, and there is a picture of what the finished product should be.

## 6.3 TYPES OF WIRING DIAGRAMS

### 6.3.1 Connection Diagrams

Connection diagrams show the physical connections, wire by wire, to devices and equipment. They also show terminal blocks, fuse blocks, resistors, lights, and so on. The connection diagram is primarily an instrument of manufacture or original installation. While it has some value to the user as a record of the general physical arrangement of the connections of an assembly, including the assembly, it is not intended nor adaptable as a means of tracing circuits through various device elements. Such tracing of circuits is nearly always accomplished more readily and accurately from the schematic or elementary diagram.

The scope of the wiring diagram is limited to the assembly under consideration and the component devices thereof. It includes neither the associated apparatus physically detached from the assembly nor the interconnecting conductors between such apparatus and the assembly.

The connection diagram, therefore, we find to be of limited value to the user after initial installation. Its chief value is found if and when it becomes necessary for the user to modify the connections of the equipment in some way. See Figure 6.1.

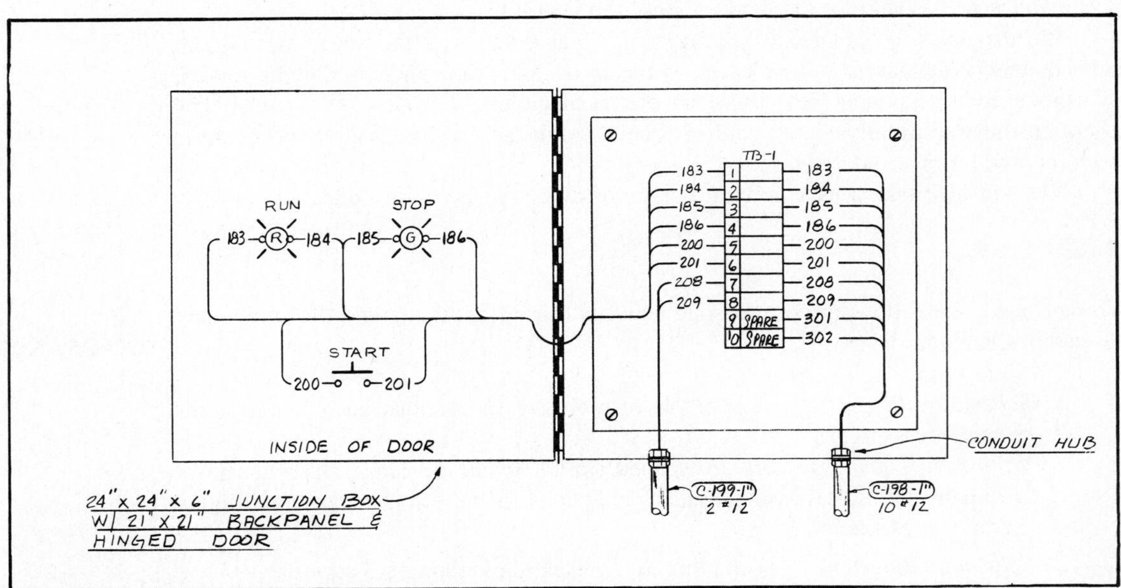


**FIGURE 6.1.** Connection Diagram

### 6.3.2 Interconnection Diagrams

Interconnection diagrams show the connections between two devices or assemblies. They also have all the characteristics of the connection diagram, physical relation between devices, and so on, but will only show those connections that are necessary to interconnect the two assemblies. Wires, complete with wire numbers, conduit numbers and terminal numbers, are shown in the interconnection diagram. See Figure 6.2.

### 6.3.3 External Wiring Diagrams

Some devices and equipment come from the manufacturer completely internally wired, and the only connections necessary are those that come from an external source into the terminals of the assembly. A drawing showing these connections is called an external wiring diagram. It shows none of the internal or factory wiring of the devices, only the external connections.

### 6.3.4 Combined Diagrams

In some cases it is convenient to combine connection and interconnection diagrams into one. This is especially true of equipment that has little or no associated apparatus. In a combined diagram, connections to associated apparatus are merely added to the connection diagram.

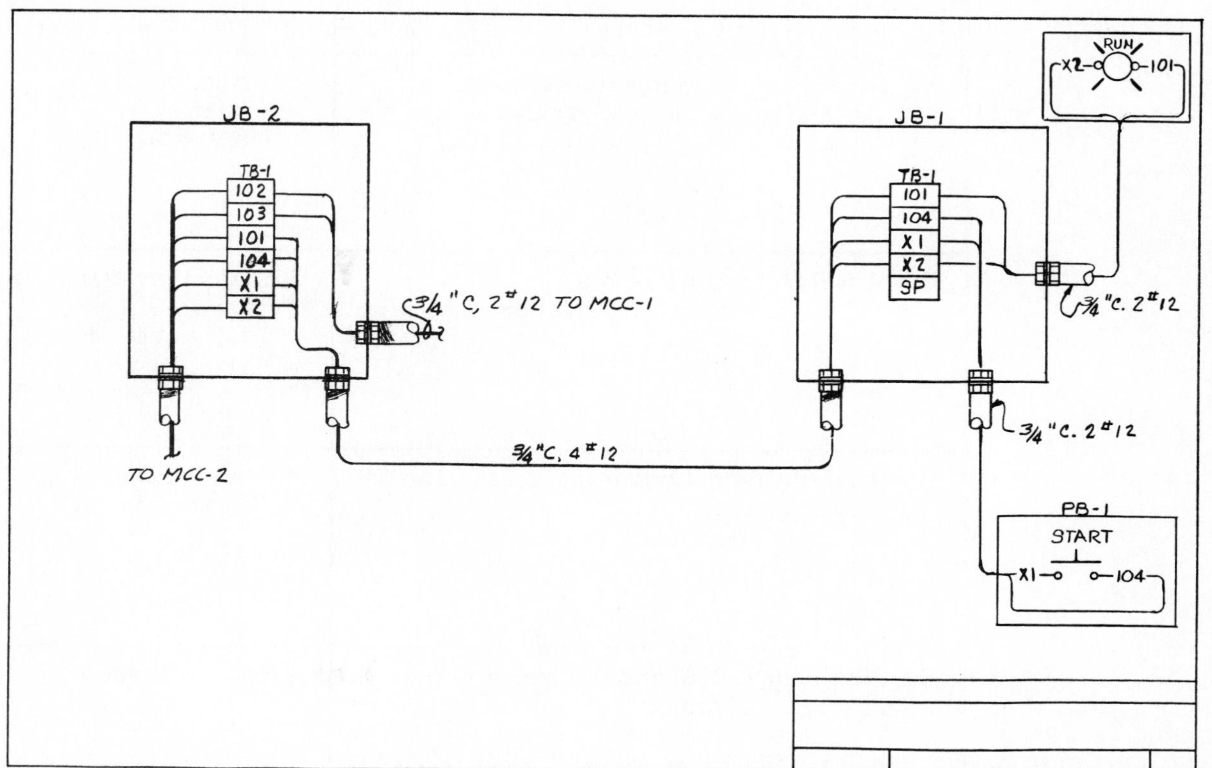


FIGURE 6.2. Interconnection Diagram

**SELF-CHECK QUIZ 6.1** **Cover the right side of the page and answer the questions.**

|  |  |
|---|---|
| 6.1 Wiring diagrams show a line for every ________. | 6.1 Ans.: *wire* Ref. 6.2 |
| 6.2 Wiring diagrams attempt to show the device's ________relationship. | 6.2 Ans.: *physical* Ref. 6.2 |
| 6.3 A diagram that shows physical connections, wire by wire, is called a ________ ________. | 6.3 Ans.: *connection diagram* Ref. 6.3.1 |
| 6.4 A diagram that shows connections between two or more devices is an ________ ________ ________. | 6.4 Ans.: *inter-connection diagram* Ref. 6.3.2 |
| 6.5 An enclosure containing terminal blocks, relays, splices, or devices is a ________ ________. | 6.5 Ans.: *junction box* Ref. 6.4.1 |
| 6.6 A NEMA type 1 box is a ________-________ box. | 6.6 Ans.: *general-purpose* Ref. 6.4.6 |
| 6.7 A NEMA type 4 box is ________and ________. | 6.7 Ans.: *water-tight* and *dust tight* Ref. 6.4.6 |
| 6.8 NEMA type 7, 8, and 9 boxes are for ________areas. | 6.8 Ans.: *Hazardous* Ref. 6.4.6 |

### 6.3.5 Field Wiring Diagrams

To further aid the field crews responsible for installation, a drawing known as a field wiring diagram is sometimes drawn. This drawing, usually reserved for instrumentation runs and connections, will show all electrical devices in their relative physical positions, although not necessarily to scale, the conduit run to each device complete with conduit number size and fill, and the individual wiring to each device, instrument, or piece of equipment shown, indicating terminal and wire numbers. This diagram can be a valuable maintenance tool. In the event there are a large number of devices or instruments, it is sometimes desirable for clarity to make field wiring diagrams by areas.

The field wiring diagram will serve the following purposes:

1. Show devices in their relative position.
2. Show all conduits associated with the devices, their size and fill.
3. Show all individual terminations complete with terminal and wire number.
4. Provide an excellent tool for maintenance and troubleshooting.
5. Provide a means of checking the area for complete device requirements and installation.

## 6.4 JUNCTION BOXES

### 6.4.1 Definition

A junction box (Figure 6.3) serves as a collection point, a termination or pull point for various wires and cables, and may contain terminal blocks, devices, relays, and/or wireway. Construction may be of steel, aluminum, plastic or fiberglass with hinged and/or bolted covers.

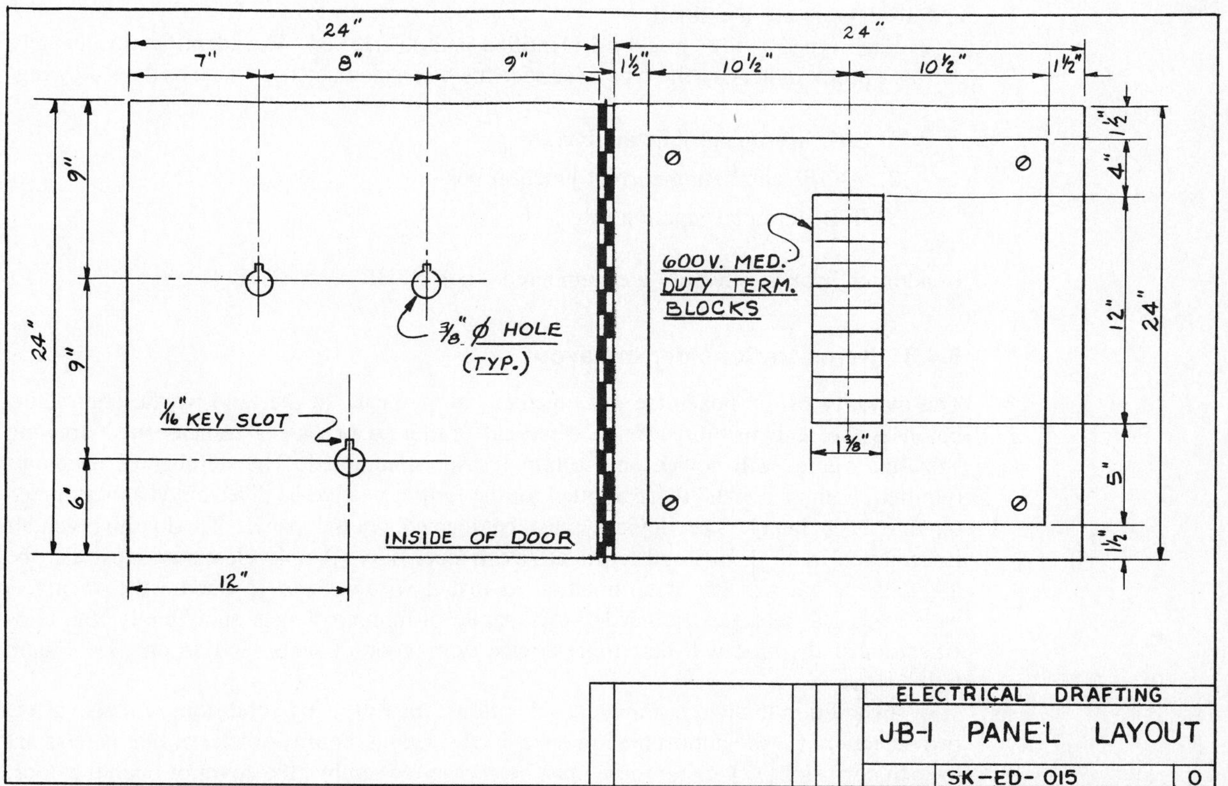


**FIGURE 6.3.** Junction Box Diagram

Area classification will generally dictate the type of box used with any special options necessary or required. Boxes located in nonclassified areas generally have no restrictions, except maybe moisture or dust resistance, and generally no restrictions on the devices mounted in or on the box, provided sufficient cubic inch area of the box and voltage requirements are met.

In a classified area the gas or vapor present in the atmosphere will spell out the conditions for certain types of boxes, devices used, sealing requirements, purging, and box construction.

### 6.4.2 Types of Boxes

Boxes fall in, basically, three major categories: (1) through or pull box, (2) terminal box or junction box, and (3) terminal and/or device box. The exterior or shape of the box is the same for all categories. It is the contents that determine the type of box.

In through or pull boxes, there are no devices or terminal blocks and the major function is to provide a pull point in a cable or wire run or a splice point. The wiring in the box may or may not be bundled and tied.

The terminal or junction box is exactly what the name implies. It usually contains terminal blocks for individual wire terminations. It may contain wireways to consolidate the wiring runs, or the wires may be bundled. Each terminal block is identified with a wire number for maintenance and troubleshooting. There may be barriers or separators depending on the voltages present in the box. Some or all of these conditions are generally associated with the terminal or junction box.

The terminal and/or device box, in addition to terminal strips, will contain some type of device, such as relays, timers, control instruments, receptacles, and switches. These devices are usually mounted to the box or on a removable back panel that is bolted to standoffs in the box. As in the junction box, all wires are identified with a wire number and all devices are identified.

Each type of box should be identified and so marked. The identification usually carries a prefix indicating the type of service the box is dedicated to, such as the following:

1. IJB: instrument junction box.
2. ACJB: alternating-current junction box.
3. TJB: terminal junction box.

In addition, boxes generally are numbered, such as IJB-1 or ACJB-1.

### 6.4.3 Items to Consider in Layout

The majority of the boxes are put together, so to speak, in the field by the contractor, but it is the responsibility of the electrical drafter to prepare a dimensioned drawing showing where each device and terminal strip is mounted. The number of terminals required, both active and those allotted for the future, need to be determined and mounted on one back panel. The devices must be located on the panel. To do this, vendor information as to device and terminal size is necessary. Ample clearance must also be allowed for wiring. The determination to install wireways or to bundle the wiring is necessary. Barriers may be needed to separate different voltages and, finally, the classification of the area will determine device type, conduit seals, and so on. See Figure 6.4.

In addition to the preceding considerations, the following conditions will also affect box selection: (1) the atmosphere may be corrosive and require a material that is resistant to such corrosion, (2) ample room must be allowed to remove the cover or open the door, and (3) the type of mounting (floor, wall, or rack) must be considered.

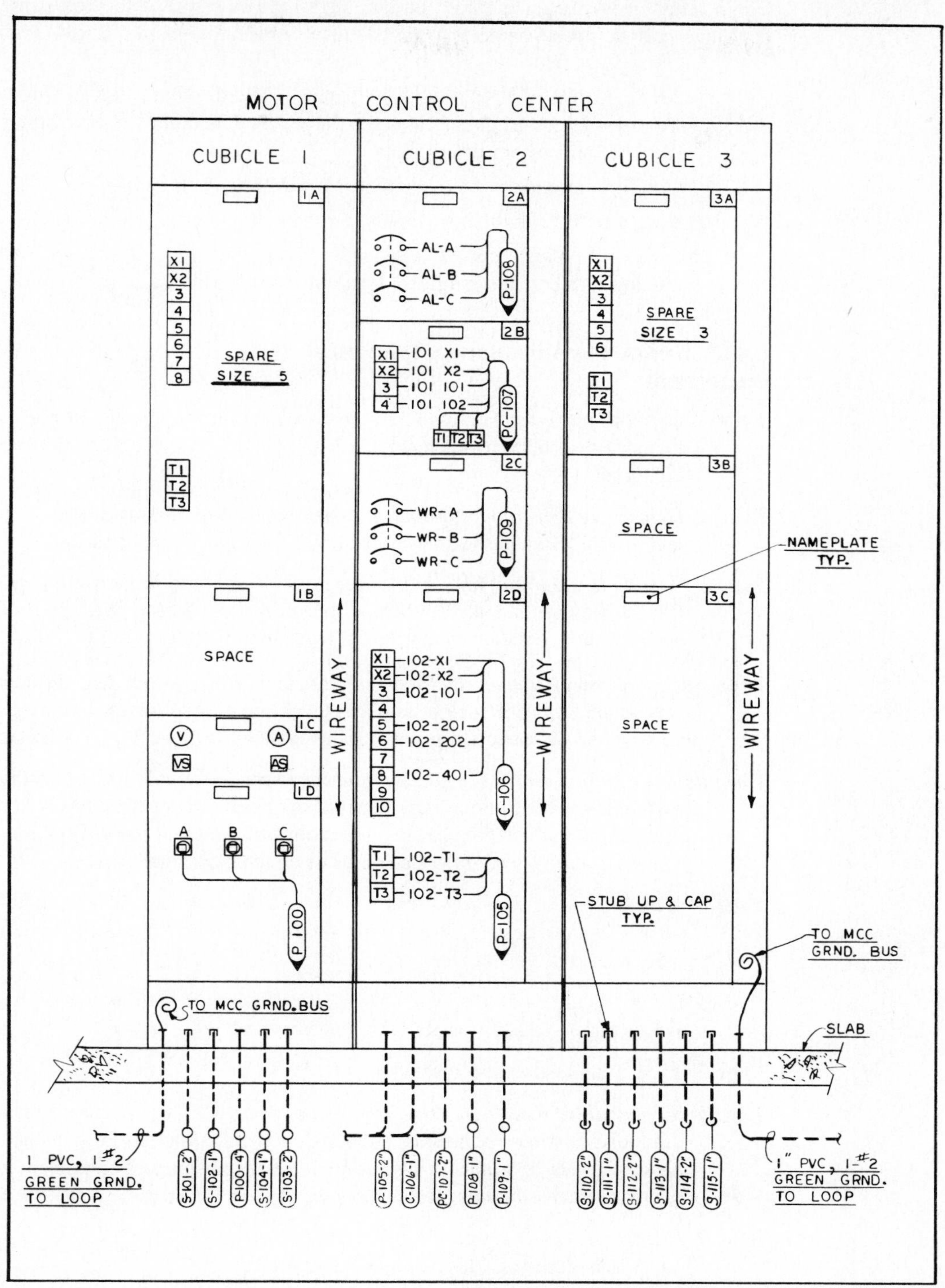


**FIGURE 6.4.** External Connection Diagram

### 6.4.4 Drawings Generated

When junction boxes are required on a project, one or more of the following types of drawings may be necessary to provide adequate information to the contractor:

1. Panel layouts to show the location of all devices and terminal strips.
2. Point-to-point wiring diagrams showing each wire complete with number, where it starts, and where it terminates.
3. Conduit entrance to box, templates for drilling, and details.
4. Mounting details to show how the box is supported and on what.
5. If several boxes are used, a schedule giving pertinent information such as size, catalog number, box number, location, and area on each box may be required.

### 6.4.5 NEMA Classifications of Electrical Equipment

**Type 1** *General purpose:* For indoor use where oil, dust, or water is not a problem. Primarily intended to prevent accidental contact by personnel with the enclosed equipment.

**Type 2** *Drip proof/drip tight:* Similar to type 1 with drip shields added to protect the enclosed equipment against falling, noncorrosive liquids and dirt.

**Type 3** *Dust tight, raintight, and sleet and ice resistant:* For outdoor use. Protects against windblown dust and water; not dust and snow proof. Has provisions for watertight connection in conduit entrance and provisions for locking.

**Type 3R** *Rain resistant and sleet and ice resistant:* For outdoor use. Protects against rain, prevents entrance of rain at a level higher than lowest live part. Not dust, snow, or sleet and ice proof. Has provisions for drainage and locking.

**Type 3S** *Dust tight, raintight, and sleet and ice proof:* For outdoor use. Protects against windblown dust and water, permits operation when enclosure is covered by external ice, does not protect equipment from internal icing. Has provisions for locking and watertight connections at conduit entrance.

**Type 4** *Watertight and dust tight:* For indoor and outdoor use. Protects against water splash, water seepage, water from hoses, and severe external condensation. Is sleet resistant but not sleet proof.

**Type 4X** *Watertight, dust tight, and corrosion resistant:* Has same provisions as type 4 enclosures; in addition, is corrosion resistant.

**Type 5** Type 5 enclosures are deleted.

**Type 6** *Submersible, watertight, dust tight, and sleet and ice resistant:* For indoor and outdoor use where occasional submersion is encountered. Protects against static head of 6 feet of water for 30 minutes, dust, splashing or external condensation of noncorrosive liquids, water, and seepage.

**Type 7** *Hazardous areas (groups A through D):* For indoor class I locations containing hazardous gases or vapors; air-break contacts.

**Type 8** *Hazardous areas (groups A through D):* For indoor class I locations containing hazardous gases or vapors; oil-immersed contacts.

**Type 9** *Hazardous areas (groups E, F, G):* For class II locations containing hazardous dusts. Prevents entrance of explosive amounts of hazardous dust.

**Type 10** *Bureau of Mines:* Designed to meet requirements of U.S. Bureau of Mines; suitable for application in coal mines.

**Type 11** *Industrial, dust tight and drip tight:* For indoor use. Protects against fibers, lint, dust, and dirt and light splashing, dripping, and external condensation of noncorrosive liquids. Has no holes, conduit knockouts, or openings, except when provision is made for oil-tight or dust-tight mechanisms with oil-resistant gaskets.

### 6.4.6 NEMA Definitions

*Acid-resistant:* Constructed so that it will not be injured readily by exposure to acid fumes.

*Dustproof:* Constructed or protected so that dust will not interfere with its successful operation.

*Dust tight:* Constructed so that dust will not enter the enclosure case.

*Fume resistant:* Constructed so that it will not be injured readily by exposure to the specified fumes.

*Moisture resistant:* Constructed or treated so that it will not be injured readily by exposure to a moist atmosphere.

*Oil tight:* Constructed so that oil will not enter the enclosure case.

*Raintight:* Constructed or protected so that exposure to a beating rain will not result in the entrance of water.

*Sleetproof:* Constructed or protected so that the accumulation of sleet will not interfere with its successful operation.

*Splashproof:* Constructed and protected so that external splashing will not interfere with its successful operation.

*Submersible:* Constructed so that it will operate successfully when submerged in water under specified conditions of time and pressure.

*Watertight:* Provided with an enclosing case that will exclude water applied in the form of a hose stream under specified conditions.

*Weatherproof (outside exposure):* Constructed or protected so that exposure to the weather will not interfere with its successful operation.

### 6.4.7 IEC Classes for Enclosures

Enclosure classification consists of the two letters IP and two numerals that describe the protection required (see Table 6.1). The first numeral designates protection of persons against contact with live or moving parts inside the enclosures and protection of equipment against the entrance of solid foreign objects. The second numeral designates protection of equipment against the entrance of water.

The following gives the various types of IEC enclosures for protective electrical equipment against explosion.

*(EX)E: Increased safety design*. Construction used for equipment that does not normally produce sparks (e.g., transformers, motors without slip rings and commutators, luminaires).

**Table 6.1 MOST FREQUENTLY USED ENCLOSURES (IEC)**

| Characteristic Letters | First Numeral | | Second Numeral 0 | 1 | 2 | 3 | 4 | 5 | 6 | 7 | 8 |
|---|---|---|---|---|---|---|---|---|---|---|---|
| | 0 | Motors[a]<br>Enclosures[b] | IP00 | — | — | — | — | — | — | — | — |
| | 1 | Motors<br>Enclosures | IP10<br><br>1 | IP11<br>ODP<br>2 | IP12<br><br>3 | — | — | — | — | — | — |
| | 2 | Motors<br>Enclosures | IP20<br><br>1 | IP21<br><br>2 | IP22<br>DPFG<br>3 | IP22<br>OSP<br>3R | IP23<br>WPI | — | — | — | — |
| IP[c]<br>(International protection) | 3 | Motors<br>Enclosures | IP30<br><br>1 | IP31<br>DPSG<br>2 | IP32<br>DPFG<br>3 | IP33<br>WPII<br>3R | IP34<br>WPII<br>3S | — | — | — | — |
| | 4 | Motors<br>Enclosures | IP40<br><br>1 | IP41<br>WPII<br>2 | IP42<br>WPII<br>3 | IP43<br>WPII<br>3R | IP44<br>WPII<br>3S | — | — | — | — |
| | 5 | Motors<br>Enclosures | IP50<br>DUIP | — | — | — | IP54<br>TEFC<br>3S | IP55<br>TEWP<br>4 | — | — | — |
| | 6 | Motors<br>Enclosures | IP60<br>DUIP<br>12 | — | — | — | — | IP65<br>TENV<br>4 | IP66<br>TEEW<br>4X | IP67<br>TEEW<br>6 | IP68 |

[a]*Closest NEMA equivalent to IEC for motor enclosures.*

[b]*Closest NEMA equivalent to IEC for all other enclosures.*

[c]*Add R after IP for pipe-ventilated machines; add W after IP for weather-protected machines.*

*(EX)D: Flameproof enclosures.* Construction that houses the components that produce arcs and sparks. During operation, interior protected from explosive gas–air mixtures from exterior; internal explosion does not ignite external gases.

*(EX)F: Pressurized enclosure.* Inside of enclosure pressurized with fresh air or inert gas at higher pressure than outside atmosphere to prevent external explosion from reaching interior.

*(EX)I: Intrinsically safe construction.* Construction that prevents internal explosions by limiting current and voltage as a function of inductance, capacitance, and resistance; power rating of circuit is smaller than the minimum ignition power of 20 microwatts. Used for signaling, metering, and automatic control circuits.

*(EX)S: Special construction.* All other types of protection not covered in the preceding will be determined by the job requirements.

# 7 LIGHTING

## 7.1 GENERAL

Lighting drawings generally fall into two broad categories: (1) area lighting, and (2) local lighting. Each will be complete in itself, showing fixture selection, mounting, location of fixtures, circuitry, conduit runs, panel locations, and connections. In addition, a list of general notes and reference drawings is included to aid in the installation.

Lighting design is a field in itself and requires a broad knowledge of methods and measurements. In this chapter we will look at some basic definitions, make a simple lighting layout, and examine the characteristics of a good lighting plan and detail sheet.

As an electrical drafter your responsibility will be to record the thinking of the design group in such a manner that the worker in the field can make the installation with a minimum of time and material. To do this, it is necessary for you to know the basic symbols and definitions used in lighting drawings.

## 7.2 BASIC DEFINITIONS AND DESCRIPTIONS

*Incandescent-filament lamp:* A lamp in which light is produced by a filament heated to incandescence by an electric current.

*Mercury-vapor lamp:* A lamp in which the active gas is mercury vapor.

*Fluorescent:* A low-pressure mercury electric discharge lamp in which a fluoresceing coating (phosphor) transforms some of the ultraviolet energy generated by the discharge into light.

*Lighting panel:* A means of grouping fuses or circuit breakers into one housing for control of general-purpose lighting and receptacle branch circuits.

*Fixture:* A lighting device, including housing lamp, socket, and sometimes reflector.

*Flush mount:* Recessed in wall or ceiling so that the face of panel or fixture is flush with the wall or ceiling.

*Surface mount:* Mounted on the wall or ceiling surface.

*Emergency light:* Light that comes on in the event of power failure; generally operated by a battery or from an auxiliary power source.

*Night light:* Fixture that is on and remains on after all other lights are turned out.

*Switched:* Can be de-energized by a switch.

*Unswitched:* Has no switch in the circuit.

*Stanchion mount:* Mounted on a pipe or stanchion; generally refers to a light fixture.

*Wall mount:* Mounted on a vertical wall; generally refers to a light fixture.

## 7.3 SYMBOLS FOR LIGHTING LAYOUTS (AMERICAN NATIONAL STANDARDS INSTITUTE)

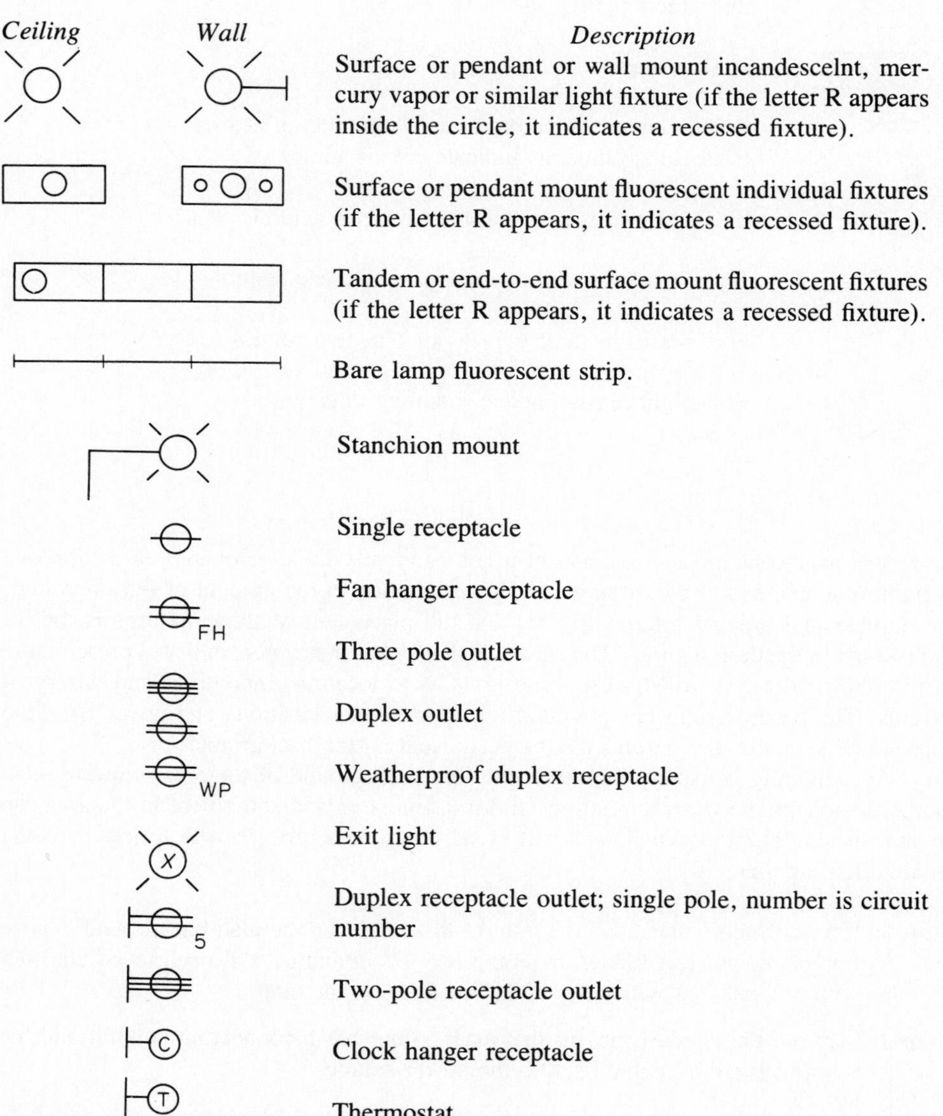

*Switch Outlets*

| | |
|---|---|
| S | Single-pole switch |
| $S_2$ | Double-pole switch |
| $S_3$ | Three-way switch |
| $S_4$ | Four-way switch |
| $S_K$ | Key-operated switch |
| $S_P$ | Switch and pilot light |
| $S_A$ | Single-pole switch; letter indicates fixtures controlled by switch. |
| (E) | Emergency lighting |
| (E) 1A | Fluorescent fixture: E = emergency, N = night light, 1 = circuit number, A = controlled by SW W/some letter |
| | Lighting panel |
| | Receptacle panel |
| | Power panel |
| 1 2 | Branch circuit home run to panel; number of arrows = number of circuits; numerals indicate circuit number(s) |
| | Indicates two wires: one hot 1 , one neutral 2 |
| | Three marks = three wires: two hot, one neutral |
| | Four marks = four wires, two hot, two neutral |
| | Floodlight; arrow indicates aiming direction |

## 7.4 AREA LIGHTING

Area lighting means the illumination of a comparatively large or open area as opposed to lighting a confined or closed area. The determination of the amount of light required, the number and type of fixtures to use, and the placement of these fixtures is the responsibility of the design group. The electrical drafter has the responsibility of presentation so that the field crews will have no questions as to location, mounting, and wiring of fixtures. The design group has provided the approximate locations and now, using the appropriate symbols, the fixtures can be designated on the background.

Area lighting is usually imposed on a light background of the area in question so that a relationship between the fixtures and the structures and equipment in the area can be established. The following steps will broadly outline a procedure to follow in doing an area lighting plan:

**Step 1:** Using symbols, locate all the fixtures to be used on the plan background. Those fixtures *completely* hidden by equipment or structures will need to be clarified with a detail, but will show as dashed lines on the plan.

**Step 2:** Lay out the conduit runs so that each fixture has a connecting conduit and the whole system is routed back to the power source.

**Step 3:** Circuiting the system will show the number of wires necessary in each conduit, and with this information the conduit can be sized.

**SELF-CHECK QUIZ 7.1** **Cover the right side of the page and answer the questions.**

|  |  |
|---|---|
| 7.1 There are generally two broad categories of lighting, ________ and ________. | 7.1 Ans.: *area* and *local* Ref. 7.1 |
| 7.2 Fuses or circuit breakers grouped into one housing for control of lighting branch circuits are called a ________ ________. | 7.2 Ans.: *lighting panel* Ref. 7.2 |
| 7.3 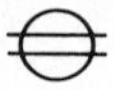 [symbol] represents ________ ________. | 7.3 Ans.: *duplex outlet* Ref. 7.3 |
| 7.4  [symbol] represents a ________. | 7.4 Ans.: *thermostat* Ref. 7.3 |
| 7.5 $S_3$ represents a ________-________ ________. | 7.5 Ans.: *three-way switch* Ref. 7.3 |
| 7.6 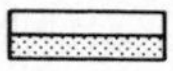 [symbol] represents a ________ ________. | 7.6 Ans.: *lighting panel* Ref. 7.3 |
| 7.7  [symbol] represents an ________ ________. | 7.7 Ans.: *emer-gency lighting* Ref. 7.3 |
| 7.8 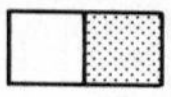 [symbol] represents a ________ ________. | 7.8 Ans.: *receptacle panel* Ref. 7.3 |

**Step 4:** Provide support for all fixtures and conduit runs either through additional detail or general notes.

Let's follow this procedure while looking at Figure 10.13.

**Step 1:** The lighting fixtures have been located on this drawing using symbols to represent the different types. An explanation of these symbols and their meaning is depicted in the Fixture Schedule in Figure 10.15. We were told this by general note 2 in Figure 10.13. In keeping with the definition of an area lighting plan, this drawing shows only the lighting and its associated conduit runs that are in the area or outside the building. Local lighting or lighting in the building is covered in Figure 10.15 and will be discussed later in the text.

All the lighting shown, with the exception of the floodlights on the poles, is 120-V lighting. The service for this will come from the 120-V lighting panel in the MCC building. The floodlights are 480-V and will be serviced from a circuit breaker in the motor control center in the building.

Four types of light fixtures are shown on the plan:

(A) Indicates a high-pressure sodium fixture ceiling mount

(B) Indicates an incandescent fixture ceiling mount

(C) Indicates a high-pressure sodium fixture, except stanchion mounted

Indicates a 100-W high-pressure sodium floodlight

By each type of fixture there is a reference to a detail number and a drawing number for additional information.

**Step 2:** It is necessary to run conduit and wire to each of these fixtures. As an example, conduit L-105 is underground, as indicated by the dashed line. It stubs-up under the motor control center and runs from there to floodlight pole AL-1, where it stubs-up at the base of pole. From there it is extended up the pole to the fixtures.

Conduit L-106 begins under the 120-V lighting panel and runs underground to the pipe rack at column D-3. From this point it is extended up to the cable tray and from that point to column line B-1. The cable only is run in the cable tray. At column line B-1, the cable again enters conduit and runs to a contactor at column A-1.

A contactor is a device that is activated by an external device, in our case a photocell, and can be used to switch a large number of lights at predetermined times. From this contactor, conduits are run to all the fixtures under the pipe rack and at the two pump pads.

It should now be easy to trace the conduits to the fixtures under the pipe rack and to light poles AL-2 and AL-3.

**Step 3:** Our plan (Figure 10.13) shows eight 200-W fixtures mounted under or near the pipe rack. This totals 1600 W, so it is necessary to run two 120-V circuits from our contactor to the lighting panel in the building. This requires four wires, #12, so conduit L-106 will contain 1/4/C #12. We make this a multiconductor cable instead of four single wires, as part of the run is in open tray.

Each floodlight pole will require two wires, so conduits L-104 and L-107 to poles AL-1 and AL-2 will contain two #10 single conductors as the entire run is in conduit. The run to pole AL-3 should be a 2/C #10 as part of the run is in cable tray.

**Step 4:** Notes indicate details showing fixture support and conduit support as required.

## 7.5 LOCAL LIGHTING

Local lighting can be defined as the illumination of a confined or enclosed area such as a room or building. The same prerequisites apply to local lighting as to area lighting. The design group determines the fixture type, location, and quantity and passes this information on to the drafter. Local lighting is also imposed on a background, or plan view of the area to be illuminated, and the same general steps for developing this drawing apply.

Let's look at Figure 10.15, a lighting and receptacle plan of the interior of the MCC building. It shows the symbols for the fixtures, fluorescent in this case, the fixtures outside the doors, the conduit runs, and the wire in the conduits. It also shows the switches at the door. The receptacle plan shows the location of the receptacles and their conduit runs from lighting panel LP-1. The fixture schedule describes each type of fixture. All this lighting is 120 V and originates at the lighting panel. General notes provide additional information to aid in the installation or to cover special conditions that exist.

## 7.6 CONVENIENCE OUTLETS AND RECEPTACLES

In addition to lighting, the lighting plans sometimes show receptacles and small power requirements, such as drinking fountains, fans, and heaters, that are in the 120–240 Vac range. This is done provided it will not cause an undue amount of congestion on the plan. If it is too crowded, the receptacle plan can be made a separate drawing, as was done with our package.

## 7.7 CIRCUITING LIGHTING LAYOUTS

At some point in the design process, a determination was made as to the voltage and type of system to be used for lighting. These data will be part of the one-line diagram and will be one of the following:

1. 480Y/277 V, 3ϕ, four wire
2. 208Y/120 V, 3ϕ, four wire
3. 120/240 V, 1ϕ, three wire

Large areas to be lighted often use the 480Y/277 V, 3ϕ, four wire system with the fixtures rated for 277-V service. This system is ideal for large lighting loads and motor loads being supplied 480 V, 3ϕ, because the one system does the whole job. Any 120-V, 1ϕ service required must come from another source.

The 208Y/120 V, 3ϕ, four-wire system is widely used because it provides lighting service at 120 V. Similarly to the 480Y/277 V system, three-branch circuits, properly balanced, can be run using only one neutral return. See Figure 7.1.

Assuming Figure 7.1 to represent a 208Y/120 V, 3ϕ, four-wire panel, we can state the following:

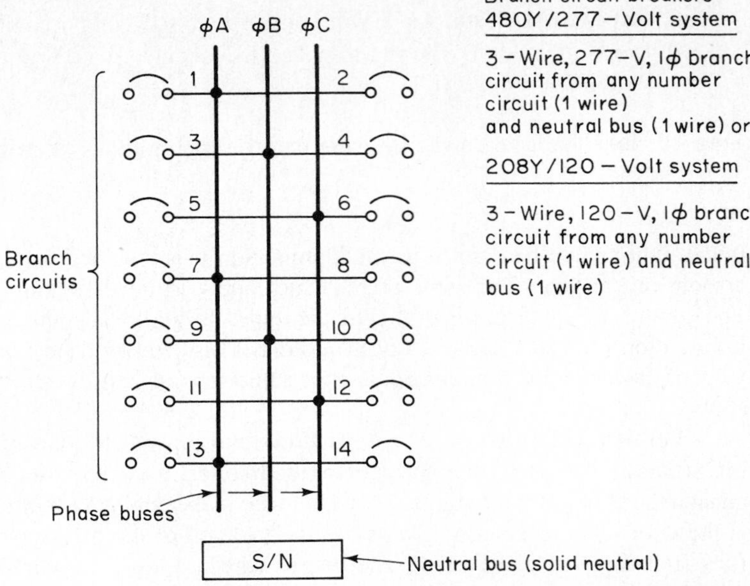


**FIGURE 7.1.** Typical 3ϕ Lighting Panel Arrangement for 3ϕ, Four-Wire Distribution and 120-V or 277-V 1ϕ Branch Circuits

1. Between any phase (A, B, or C) and the neutral bus, a voltmeter would read 120 volts.
2. Between any two phases (A and B, B and C, or A and C), a voltmeter would read 208 V.
3. We can carry as many as three circuits by only running one neutral, or a total of four wires, providing each circuit is fed from a different phase. In large installations this will effect a savings in wire and labor.
4. The panel in Figure 7.1 shows 12 circuits, but as many as 42 circuits maximum can be put in one panel housing.

The 120/240, 1ϕ, three-wire system is most commonly used in residential wiring and to some extent in commercial and industrial. Figure 7.2 shows this panel arrangement.

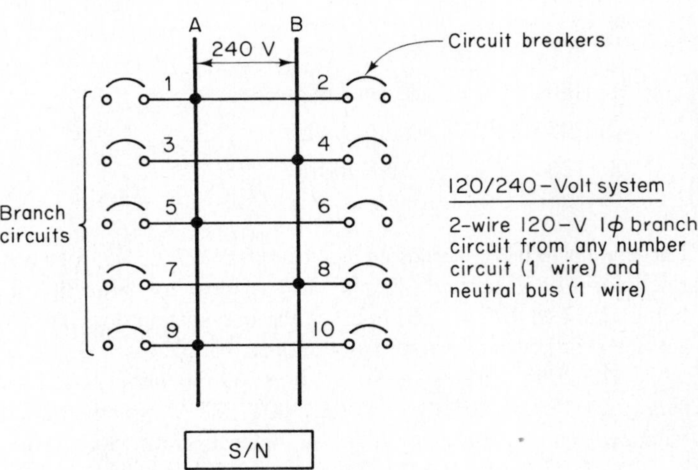


**FIGURE 7.2.** Typical Lighting Panel Arrangement for 120/240 Volt, 1ϕ Service

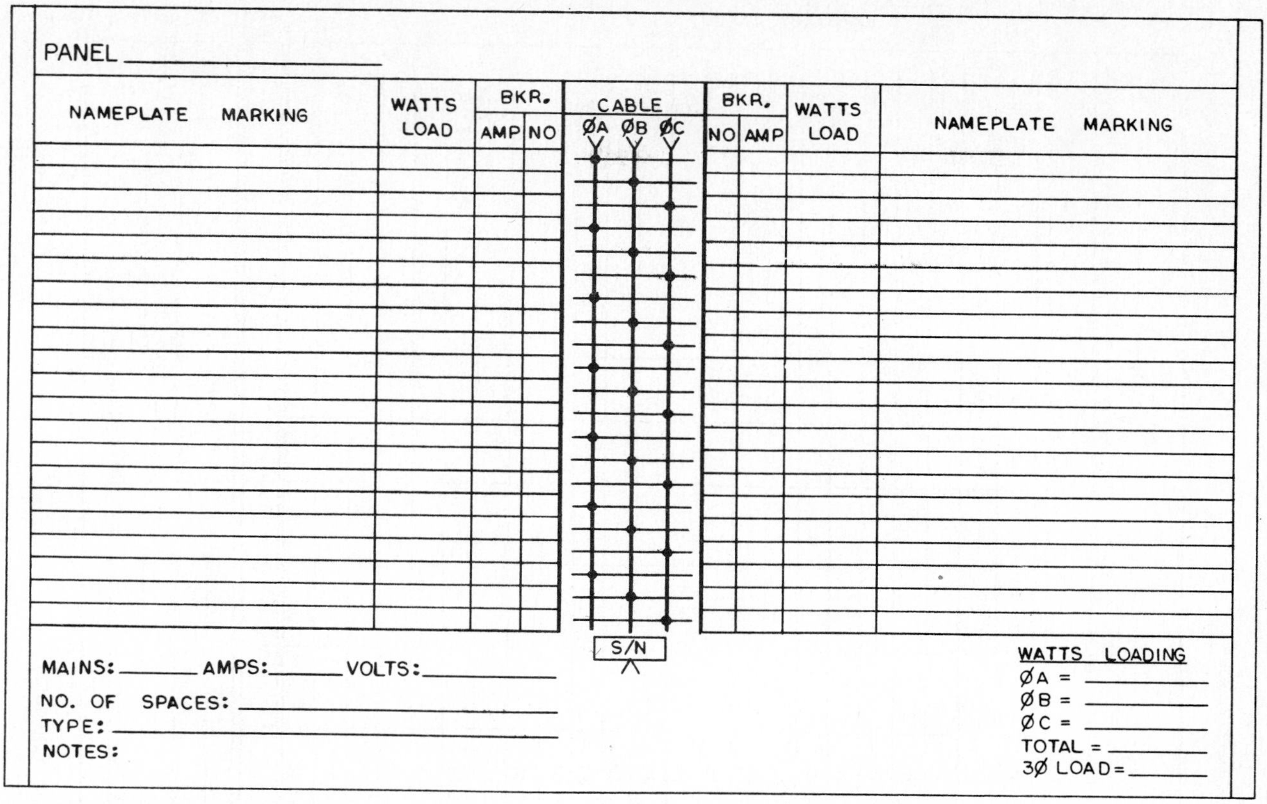

PANEL ____________

| NAMEPLATE MARKING | WATTS LOAD | BKR. | | CABLE | BKR. | | WATTS LOAD | NAMEPLATE MARKING |
|---|---|---|---|---|---|---|---|---|
| | | AMP | NO | ØA ØB ØC | NO | AMP | | |

S/N

MAINS: ______ AMPS: ______ VOLTS: ____________

NO. OF SPACES: ____________

TYPE: ____________

NOTES:

WATTS LOADING

ØA = ________

ØB = ________

ØC = ________

TOTAL = ________

3Ø LOAD = ________

**FIGURE 7.3.** $3\phi$ Panel Schedule

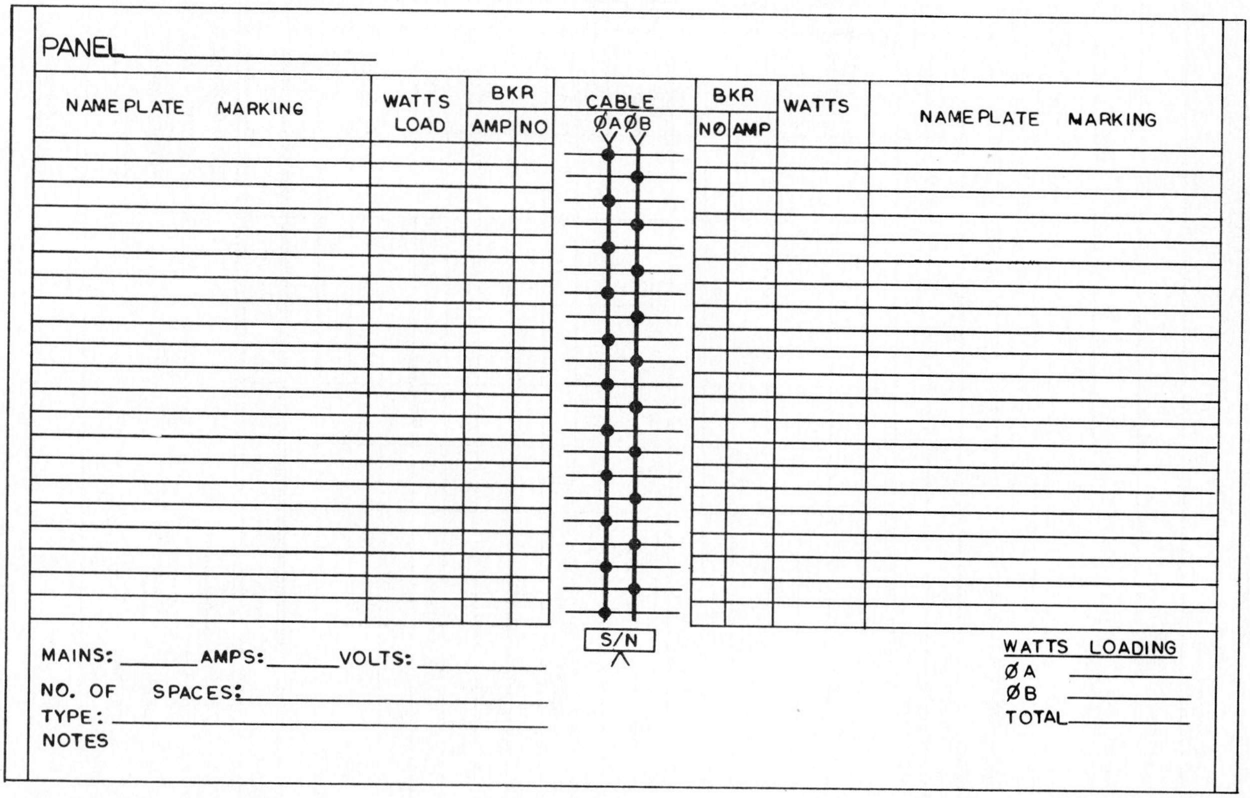

PANEL ____________

| NAMEPLATE MARKING | WATTS LOAD | BKR | | CABLE | BKR | | WATTS | NAMEPLATE MARKING |
|---|---|---|---|---|---|---|---|---|
| | | AMP | NO | ØAØB | NO | AMP | | |

S/N

MAINS: ______ AMPS: ______ VOLTS: ____________

NO. OF SPACES: ____________

TYPE: ____________

NOTES

WATTS LOADING

ØA ________

ØB ________

TOTAL ________

**FIGURE 7.4.** Panel Schedule, $1\phi$

# LIGHTING PANEL SCHEDULE

PANEL NO ______ MFG ______ LOCATION ______

SERVICE VOLTAGE ______ TYPE ______ MOUNTING ______

______ BUS RATING ______ DRAWING NO. ______

| DESCRIPTION | BREAKER | | VOLT AMPS | | | CIRCUIT | BUS | CIRCUIT | VOLT AMPS | | | BREAKER | | DESCRIPTION |
|---|---|---|---|---|---|---|---|---|---|---|---|---|---|---|
| | POLE | AMP | A | B | C | | | | A | B | C | POLE | AMP | |
| | | | | | | 1 | | 2 | | | | | | |
| | | | | | | 3 | | 4 | | | | | | |
| | | | | | | 5 | | 6 | | | | | | |
| | | | | | | 7 | | 8 | | | | | | |
| | | | | | | 9 | | 10 | | | | | | |
| | | | | | | 11 | | 12 | | | | | | |
| | | | | | | 13 | | 14 | | | | | | |
| | | | | | | 15 | | 16 | | | | | | |
| | | | | | | 17 | | 18 | | | | | | |
| | | | | | | 19 | | 20 | | | | | | |
| | | | | | | 21 | | 22 | | | | | | |
| | | | | | | 23 | | 24 | | | | | | |
| | | | | | | 25 | | 26 | | | | | | |
| | | | | | | 27 | | 28 | | | | | | |
| | | | | | | 29 | | 30 | | | | | | |
| | | | | | | 31 | | 32 | | | | | | |
| | | | | | | 33 | | 34 | | | | | | |
| | | | | | | 35 | | 36 | | | | | | |
| | | | | | | 37 | | 38 | | | | | | |
| | | | | | | 39 | | 40 | | | | | | |
| | | | | | | 41 | | 42 | | | | | | |
| TOTALS | | | | | | | | | | | | | | |

BUS A ______ MAIN (BREAKER) (LUGS) ______ LINE AMPS ______

BUS B ______ LOCATION (TOP) (BOTTOM) ______ PHASE @ ______ VOLTS

BUS C ______ FEEDER SIZE ______ KVA DEMAND ______

TOTAL ______ SOURCE ______ DATE ______

**FIGURE 7.5.** Lighting Panel Schedule

| BRANCH CIRCUIT DATA | | | | | | | | BRANCH CIRCUIT DATA | | | | PROTECTION | | |
|---|---|---|---|---|---|---|---|---|---|---|---|---|---|---|
| CIRCUIT NUMBER | CONNECTION | | CONNECTED LOAD | | | | VOLTS & PHASE | NO. & SIZE OF CONDUIT | NO. & SIZE OF CNDCT. | (1) VOLTAGE DROP | | CIRCUIT BREAKER | | |
| | FROM | TO | HP | PF | AMPS | LENGTH | | | | FACTOR (2) | % (3) | POLES | FRAME | TRIP |
| | | | | | | | | | | | | | | |
| | | | | | | | | | | | | | | |
| | | | | | | | | | | | | | | |
| | | | | | | | | | | | | | | |
| | | | | | | | | | | | | | | |
| | | | | | | | | | | | | | | |
| | | | | | | | | | | | | | | |
| | | | | | | | | | | | | | | |
| | | | | | | | | | | | | | | |
| | | | | | | | | | | | | | | |
| | | | | | | | | | | | | | | |
| | | | | | | | | | | | | | | |
| | | | | | | | | | | | | | | |

(1) FOR VOLTAGE DROP REFER TO APPLICABLE CHART

(2) $\text{FACTOR} = \frac{\text{VOLTAGE DROP}}{\text{10,000 AMP FT}}$

(3) $\%\ \text{VOLTAGE DROP} = \frac{\text{FACTOR} \times \text{AMPERES} \times \text{CIRCUIT FEET} \times 10^2}{\text{VOLTAGE (PHASE TO PHASE)}}$

**FIGURE 7.6.** Panel Schedule

From Figure 7.2, we can make the following statements:

1. Between either bus A or B and the neutral bus, a voltmeter would read 120 V.
2. Between bus A and bus B, a voltmeter would read 240 V.
3. From this panel we can only carry two circuits with a common neutral, providing they are fed from different phases.

## 7.8 PANEL SCHEDULES

Figure 10.15 shows one type of panel schedule used to identify what each circuit breaker controls. Other types or variations of panel schedules are shown in Figures 7.3 through 7.6.

# POWER

In any project, especially an industrial facility, various pieces of machinery and equipment require electric power before they can function and fulfill an intended use. The power to operate these devices is delivered via wire in conduit, cable tray, duct, or some form of raceway. The method in which this power is delivered, the routing it takes, the means of support, sizes of wire, conduit, cable tray, and its termination are all part of the *power plan*. By definition then, "A power plan is a plan drawing showing all the power requirements of the area covered by the background."

## 8.1 GENERAL

Power plans, like grounding and lighting plans, are imposed on a background of the area in question so that a relationship can be established. The background should be lighter than the power symbols used so that the power portion of the drawing will stand out. Figure 10.12 is a typical power plan.

There may be times when the area is so congested or so large it may require more than a single drawing. This is perfectly acceptable, for if it takes two or more drawings to accurately and clearly show the installation, the extra drawing(s) are warranted.

## 8.2 SCOPE

It will be necessary to recognize the symbols used to understand what is taking place.

## 8.3 PLAN SYMBOLS: POWER

------------o

The dashed line indicates concealed conduit, in this case buried or underground conduit. The circle is the point at which the conduit turns up and continues above grade toward the viewer.

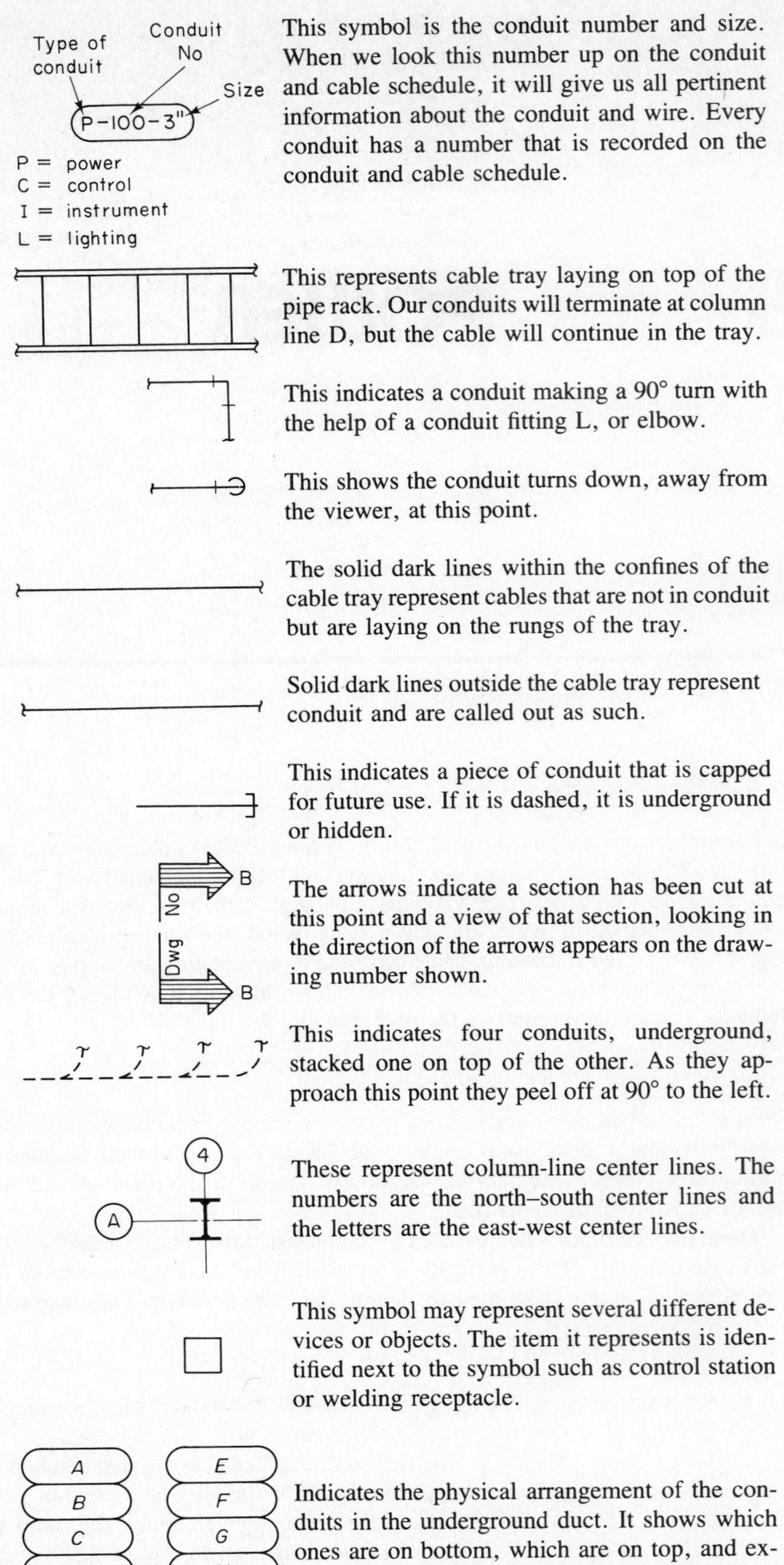


This symbol is the conduit number and size. When we look this number up on the conduit and cable schedule, it will give us all pertinent information about the conduit and wire. Every conduit has a number that is recorded on the conduit and cable schedule.

This represents cable tray laying on top of the pipe rack. Our conduits will terminate at column line D, but the cable will continue in the tray.

This indicates a conduit making a 90° turn with the help of a conduit fitting L, or elbow.

This shows the conduit turns down, away from the viewer, at this point.

The solid dark lines within the confines of the cable tray represent cables that are not in conduit but are laying on the rungs of the tray.

Solid dark lines outside the cable tray represent conduit and are called out as such.

This indicates a piece of conduit that is capped for future use. If it is dashed, it is underground or hidden.

The arrows indicate a section has been cut at this point and a view of that section, looking in the direction of the arrows appears on the drawing number shown.

This indicates four conduits, underground, stacked one on top of the other. As they approach this point they peel off at 90° to the left.

These represent column-line center lines. The numbers are the north–south center lines and the letters are the east-west center lines.

This symbol may represent several different devices or objects. The item it represents is identified next to the symbol such as control station or welding receptacle.

Indicates the physical arrangement of the conduits in the underground duct. It shows which ones are on bottom, which are on top, and exactly how the conduits are stacked in the trench.

## 8.4 READING THE POWER PLAN

With an understanding of the basic symbols used, it is much easier to follow the conduit and cable runs on the power plan, although a complete understanding of the plan is difficult without the aid of other related drawings. These drawings are listed on the right side of the sheet under the heading Reference Drawings. Here is found a list of the drawings necessary to furnish complete information on the power portion of the project.

Special conditions and additional information are found on the upper right side of the sheet under the heading General Notes.

Let's follow one conduit in Figure 10.12 from beginning to end. Conduit PC-112-1" appears on our power plan. By examining the power plan, we see it goes to the sump pump P-102. Referring to the conduit and cable schedule (Figure 10.9) and finding conduit PC-112, we obtain the following information:

1. It is a conduit containing both power and control wiring. The prefix PC tells this.
2. It begins in MCC-1 and goes to the motor connection box on motor P-102.
3. Part of the run is in cable tray (see remarks).
4. It is a 1-in. PVC conduit and contains 1-3/C #12 and 1-5/C #12, which are multiconductor cables, one cable containing three #12 wires and one with five #12 wires.

Knowing the conduit comes from MCC-1 and knowing several conduits stub-up underneath MCC-1, it is necessary to determine which is PC-112. This information can be found in Figure 10.16. It shows PC-112 to be in the row of conduit stub-ups in the back of the MCC, the third conduit from the north end. The power plan (Figure 10.12) shows this underground duct bank leaving the MCC building. There is a section cut with a reference to drawing SK-E-119 (Figure 10.24). We find this section, B-B, in the lower left-hand corner. This section shows conduit PC-112 to be the top conduit on the right side.

With this information we can follow PC-112 out of the MCC, along the underground duct to where it turns left and stubs-up at column line D. From there the conduit stops, but the cables continue in the cable tray to a point just past column line C, where they leave the tray and enter conduit again. The conduit now goes east along column line C to the edge of the piperack, then turns down, goes underground, and heads toward P-102, where it stubs-up at the base of the pump pad. To determine how the termination is made at the motor, the note makes reference to detail A on drawing SK-E-113 (Figure 10.18). This detail, an elevation and plan, explains and depicts how the conduit is connected to the control station and the motor and describes the material required to make the connection.

## 8.5 POWER DETAILS

It is virtually impossible to show enough detail on the plan drawing for the installation to be made. Therefore, some enlarged details of some areas are usually required. These details appear on a separate drawing entitled Power Details (see Figure 10.18):

1. Detail A shows a plan view and an elevation of the conduit routing to the sump and to the control station. All items are called out and reference made to other drawings and details.
2. Detail B is also a two-view detail showing conduit terminations for the two big motors.
3. Detail C shows the material and mounting for a stanchion for the control station.
4. Detail D explains the contactor mounting location and related conduit runs.

**SELF-CHECK QUIZ 8.1** **Cover the right side of the page and answer the questions.**

| | |
|---|---|
| 8.1 A power plan is a drawing showing all the ________ requirements of the area or project. | 8.1 Ans.: *power*<br>Ref. 8.1 |
| 8.2 ——+⊃ or ——+⊗ represents a conduit turning ________ from the viewer. | 8.2 Ans.: *away*<br>Ref. 8.3 |
| 8.3 ——+○ or ——+⊙ represents a conduit turning ________ the viewer. | 8.3 Ans.: *toward*<br>Ref. 8.3 |
| 8.4 - - - - - - - represents ________ ________. | 8.4 Ans.: *hidden conduit*<br>Ref. 8.3 |
| 8.5 Power plans include ________ runs. | 8.5 Ans.: *conduit*<br>Ref. 8.1 |
| 8.6 The background should be ________ than the symbols. | 8.6 Ans.: *lighter*<br>Ref. 8.2 |
| 8.7 ┌+— represents a ________ ________. | 8.7 Ans.: *90° fitting*<br>Ref. 8.3 |
| 8.8 Draw with your head instead of your ________. | 8.8 Ans.: *pencil*<br>Ref. 8.6 |

## 8.6 PROCEDURE FOR PLAN CONSTRUCTION

The following procedure is suggested for electrical drafters to aid in doing a power plan. The steps are for the individual doing the drafting, not members of the design group.

**Step 1.** Have available or construct a clean, uncluttered background of the area in question. The line work of the background should be light so that all power items put on this background will stand out to the viewer. The power items are darker than the background.

**Step 2.** Study the background along with the mark-up made by the design group. Become familiar with structures, piperacks, substations, and so on, to enable you to more accurately record the information on the mark-up.

**Step 3.** Study the power requirements and follow out the conduit routing for power sources.

**Step 4.** Follow the routing of raceways designated by the design group and assure yourself you are basically familiar with the methods used.

**Step 5.** Begin transferring the marks from the worksheet to the finished tracing, completing one power run at a time from beginning to end.

**Step 6.** Yellow out completed runs as you transfer them.

**Step 7.** After you complete the drawing, run a print and backcheck your work against the mark-up.

The secret in doing any drawing is to know what you are drawing and why. Do not hesitate to ask questions. Make sure you understand what is being done. This is the best way to learn and take the initial steps toward design status. Draw with your head instead of just your pencil.

# 9 GROUNDING

## 9.1 GENERAL

The word *grounding* as applied to electrical drafting is generally used to cover that area of equipment grounding that is primarily for personnel protection and safety. Equipment grounding is a connection to ground from one or more of the non-current-carrying metal parts of a wiring system or of apparatus connected to the system. As used in this sense, the term *equipment* includes all such metal parts as metal conduits, metal trays, metal wireways, outlet boxes, cabinets, motor frames, lighting fixtures, and switchgear.

Equipment grounding is the responsibility of the design team and becomes the responsibility of the electrical drafter to record on the drawing.

## 9.2 REASONS FOR GROUNDING

Equipment is grounded for the following reasons:

1. To ensure against dangerous electric shock voltage exposure to any person who may come in contact with electrical equipment.
2. To provide a current-carrying path of adequate capacity to accept any fault currents that may be allowed to flow before such fault is cleared.
3. To make the electrical system function properly when a grounding condition exists.
4. To provide equipment protection in the event of a fault current until such time as the fault is cleared or the circuit is de-energized through protective devices.

## 9.3 DEFINITIONS

The following are basic definitions that will aid in the understanding of grounding systems:

*Ground:* A ground is a conducting connection, whether intentional or accidental, between an electrical circuit or equipment and earth or to some conducting body that serves in place of earth.

*Grounded:* Grounded means connected to earth or to some conducting body that serves in place of earth.

*Grounded conductor:* A system or circuit conductor that is intentionally grounded, such as a grounded neutral conductor.

*Grounding conductor:* A conductor used to connect equipment or the grounded circuit of a wiring system to a grounding electrode or electrodes.

*Ground rod:* A conductive metal rod driven in the ground to serve as a grounding electrode, usually 10 to 15 ft long and 1 in. in diameter.

*Ground loop:* A continuous ground electrode of wire encircling the area and connecting three or more ground rods.

*Ground clip:* A metal clip connected to the ground loop to enable one or more extensions of the ground loop to be extended.

*Ground well:* A concrete encasement around the ground rod to allow access to the rod to measure ground resistance.

*Ground grid:* A steel mesh or wire grid buried beneath the surface of an area to serve as a ground electrode.

*Thermoweld:* A heat welding process for joining two wires or attaching steel to wire for grounding purposes.

*Ground bus:* A metal plate to which multiple ground leads are attached.

*Bonding jumper:* A reliable conductor to assure continuity between metal parts that must be electrically connected.

## 9.4 SYMBOLS

There are a number of symbols, some standardized, that are used in grounding drawings. A familiarity with these symbols will aid you to understand the basic systems.

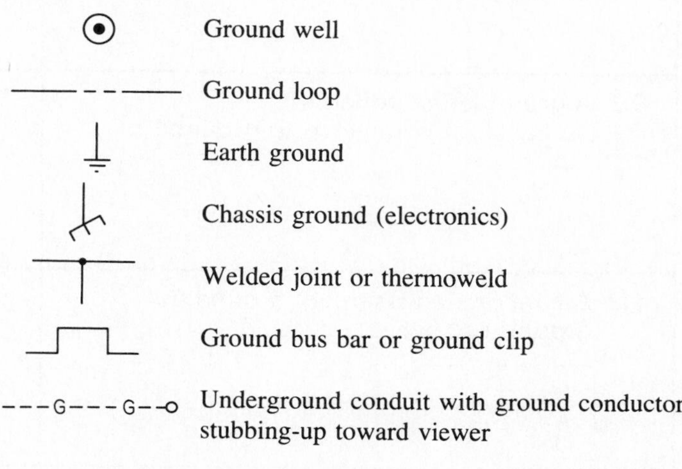

**SELF-CHECK QUIZ 9.1** **Cover the right side of the page and answer the questions.**

| | |
|---|---|
| 9.1 Grounding is primarily for ________ ________ and ________. | 9.1 Ans.: *personnel protection* and *safety* Ref. 9.1 |
| 9.2 Grounded means ________ ________ ________. | 9.2 Ans.: *connected to earth* Ref. 9.3 |
| 9.3 A conductive metal rod driven into the ground is a ________ ________. | 9.3 Ans: *ground rod or grounding electrode* Ref. 9.3 |
| 9.4 A heat welding process for joining two wires is called ________. | 9.4 Ans.: *thermoweld* Ref. 9.3 |
| 9.5 ⊙ indicates a ________ ________. | 9.5 Ans.: *ground well* Ref. 9.4 |
| 9.6 A continuous ground electrode encircling the area and connecting three or more ground rods is a ________ ________. | 9.6 Ans.: *ground loop* Ref. 9.3 |
| 9.7 A grounded conductor is ________connected to ground. | 9.7 Ans.: *intentionally* Ref. 9.3 |
| 9.8 A concrete encasement around a ground rod is a ________ ________. | 9.8 Ans.: *ground well* Ref. 9.3 |

## 9.5 NATIONAL ELECTRIC CODE ON GROUNDING

All grounding systems must be installed in accordance with the latest edition of the National Electric Code except in those areas in which the code does not apply.

Article 250 of the National Electric Code covers general requirements for grounding and bonding and specific requirements for the following:

1. Location of grounding connections.
2. Conductors to be grounded on grounded systems.
3. Types and sizes of grounding and bonding conductors and electrodes.
4. Methods of grounding and bonding.
5. Connections for lightning arrestors.

Other articles in the code, applying to particular cases of installation of conductors and equipment, contain additional requirements. These must be taken into consideration also in grounding installations.

## 9.6 LAYING OUT THE GROUNDING PLAN

The grounding plan will generally be imposed on a background of the area under consideration to be grounded. All elements of the equipment-grounding system should be darker than the background so that they stand out and there is no confusion between object lines of the background and grounding lines of the system.

Select the routing for a ground loop to encircle the area in question. This routing should be in straight lines as far as possible, with ground wells on the corners. See the typical grounding plan in Figure 10.11. After the ground loop and ground wells are located, run ground taps to all equipment and structural steel using a thermoweld process to terminate this tap at both the ground loop and the device to be grounded. It is a good practice to run the tap conductor in polyvinyl conduit (nonmetallic) for protection. Call out all ground wells or rods, equipment taps, ground loop, wire sizes, and reference to details explaining installation.

Figure 10.11 is a reasonably typical grounding plan of a process area and building. The ground loop system as opposed to a ground grid system was used in this area. A ground loop of #2/0 AWG green insulated wire encircles the entire area and also the building. Theoretically, the building and the area ground systems are two separate and distinct systems, each with its own loop, but they have been tied together for continuity. The dashed lines coming from the ground loop indicate underground, nonmetallic (PVC) conduit runs, each containing one #2 AWG green insulated wire to some structure or piece of equipment. The connection to the ground loop is by a thermoweld process. Each conduit will stub-up at a particular structure or piece of equipment. From this stub-up, the #2 wire will be extended and connected to the device or structure to be grounded. This connection is to be made with a grounding clamp. Ground clips are provided at several columns on the pipe rack and also in the building for any grounding requirements or extensions. In addition, general notes are added to clarify any special conditions and to aid in the installation.

# 10 THE DRAWING PACKAGE

**10.1 GENERAL** After the engineering phase of a project has been completed, the drawings pertaining to the project are put together to form the *package*. A typical electrical package follows. It is preceded by a drawing register that lists all the drawings in the set by drawing number and title (Figures 10.1 and 10.2). This is the set of drawings from which the crafts in the field construct the project. Each discipline will issue its own set of drawings for the project, all of which will go into the construction package.

When the construction package reaches the field crews, the structural drawings are given to the structural people, the piping section to the piping group, the electrical drawings to the electrical group, and so on. Each craft works from the set of drawings pertaining to its discipline and coordinates its work with the other crafts.

Figures 10.3 through 10.24 constitute the drawing package.

## DRAWING REGISTER

CLIENT________

PROJECT TITLE________

CLIENT JOB NO.________

MONTHLY REPORT DATES

| DRAWINGS ISSUED THIS TRANSMITTAL | DRAWING NO. | DRAWING TITLE | SCH. ISSUE | APPVL. ISSUE / RETURN | CONST. ISSUE | REVISION NO. | REVISION DATE |
|---|---|---|---|---|---|---|---|
| | SK-E-100 | CLASSIFICATION AND KEY PLAN | | | | | |
| | SK-E-101 | ONE-LINE DIAGRAM | | | | | |
| | SK-E-101A | ONE-LINE DIAGRAM | | | | | |
| | SK-E-102 | SCHEMATIC WIRING DIAGRAM SHEET 1 of 2 | | | | | |
| | SK-E-103 | SCHEMATIC WIRING DIAGRAM SHEET 2 of 2 | | | | | |
| | SK-E-103A | SCHEMATIC WIRING DIAGRAM | | | | | |
| | SK-E-104 | CONDUIT AND CABLE SCHEDULE SHEET 1 of 2 | | | | | |
| | SK-E-105 | CONDUIT AND CABLE SCHEDULE SHEET 2 of 2 | | | | | |
| | SK-E-106 | GROUNDING PLAN | | | | | |
| | SK-E-107 | POWER PLAN | | | | | |
| | SK-E-108 | AREA LIGHTING PLAN | | | | | |
| | SK-E-109 | CABLE TRAY PLAN | | | | | |

JOB NO. 0000-00 DWG. TYPE ELECTRICAL PAGE 1 of 2

## DRAWING STATUS

CLIENT________ JOB NO. 12345

DISCIPLINE ELECTRICAL

PREPARED BY E.G. SCHRIEVER

| ASSIGNED HRS. TOTAL | ASSIGNED HRS. DE | ASSIGNED HRS. DR | ASSIGNED HRS. CK | % COMPLETE 20 40 60 80 | ASSN. HR × % DE | ASSN. HR × % DR | ASSN. HR × % CK | SQ CH. ISSUE / RETURN | DESIGN CHG. HR | OTHER HRS. | REMARKS | EST. HRS. TO COMP. |
|---|---|---|---|---|---|---|---|---|---|---|---|---|
| | | | | | | | | | | | | |
| | | | | | | | | | | | | |
| | | | | | | | | | | | | |
| | | | | | | | | | | | | |
| | | | | | | | | | | | | |
| | | | | | | | | | | | | |
| | | | | | | | | | | | | |
| | | | | | | | | | | | | |
| | | | | | | | | | | | | |
| | | | | | | | | | | | | |
| | | | | | | | | | | | | |
| | | | | | | | | | | | | |
| | | | | | | | | | | | PAGE TOTALS | |

FIGURE 10.1.

## DRAWING REGISTER

CLIENT ____
PROJECT TITLE /
CLIENT JOB NO. ____

MONTHLY REPORT DATES

| DRAWINGS ISSUED THIS TRANSMITTAL | DRAWING NO. | DRAWING TITLE | SCH. ISSUE | APPVL. ISSUE / RETURN | CONST. ISSUE | REVISION NO. | REVISION DATE |
|---|---|---|---|---|---|---|---|
| | SK-E-110 | BUILDING LIGHTING AND RECEPTACLE PLAN | | | | | |
| | SK-E-111 | MCC BUILDING CONDUIT STUB-UP PLAN | | | | | |
| | SK-E-112 | GROUNDING DETAILS | | | | | |
| | SK-E-113 | POWER DETAILS | | | | | |
| | SK-E-114 | LIGHTING DETAILS | | | | | |
| | SK-E-115 | CABLE TRAY DETAILS | | | | | |
| | SK-E-116 | JB-1 WIRING DIAGRAM | | | | | |
| | SK-E-117 | FIELD WIRING DIAGRAM | | | | | |
| | SK-E-118 | MISC. DETAILS & ELEVATIONS SHEET 1 of 2 | | | | | |
| | SK-E-119 | MISC. DETAILS & ELEVATIONS SHEET 2 of 2 | | | | | |
| | | | | | | | |
| | | | | | | | |

JOB NO. 0000-00 DWG. TYPE ELECTRICAL PAGE 2 of 2

## DRAWING STATUS

CLIENT ____ JOB NO. 12345
DISCIPLINE ELECTRICIAL
PREPARED BY E. J. Schriever

| ASSIGNED HRS. TOTAL | DE | DR | CK | % COMPLETE 20 40 60 80 | ASSN. HR × % DE | DR | CK | SQ CH. ISSUE / RETURN | DESIGN CHG.HR | OTHER HRS. | REMARKS | EST. HRS. TO COMP. |
|---|---|---|---|---|---|---|---|---|---|---|---|---|
| | | | | | | | | | | | | |
| | | | | | | | | | | | | |
| | | | | | | | | | | | | |
| | | | | | | | | | | | | |
| | | | | | | | | | | | | |
| | | | | | | | | | | | | |
| | | | | | | | | | | | | |
| | | | | | | | | | | | | |
| | | | | | | | | | | | | |
| | | | | | | | | | | | | |
| | | | | | | | | | | | | |
| | | | | | | | | | | | | |
| | | | | | | | | | | | PAGE TOTALS | |

FIGURE 10.2.

GENERAL NOTES

1. FOR A COMPLETE LIST OF ELECTRICAL REFERENCE DRAWINGS SEE DRAWING NO. SK-E-100

REFERENCE DRAWINGS

| | |
|---|---|
| SK-E-100 | CLASSIFICATION & KEY PLAN |
| SK-E-101 | ONE-LINE DIAGRAM |
| SK-E-102 | SCHEMATIC WIRING DIAG. SH. 1 OF 2 |
| SK-E-103 | SCHEMATIC WIRING DIAG. SH 2 OF 2 |
| SK-E-104 | CONDUIT & CABLE SCHED. SH 1 OF 2 |
| SK-E-105 | CONDUIT & CABLE SCHED. SH 2 OF 2 |
| SK-E-106 | GROUNDING PLAN |
| SK-E-107 | POWER PLAN |
| SK-E-108 | AREA LIGHTING PLAN |
| SK-E-109 | CABLE TRAY PLAN |
| SK-E-110 | BLDG. LGT. & RECEPT. PLAN |
| SK-E-111 | BLDG. STUB-UP PLAN |
| SK-E-112 | GROUNDING DETAILS |
| SK-E-113 | POWER DETAILS |
| SK-E-114 | LIGHTING DETAILS |
| SK-E-115 | CABLE TRAY DETAILS |
| SK-E-116 | JB-1 WIRING DIAGRAM |
| SK-E-117 | FIELD WIRING DIAGRAM |
| SK-E-118 | MISC. DETAILS SH. 1 OF 2 |
| SK-E-119 | MISC. DETAILS SH 2 OF 2 |

1 2 3 4
A B C D
N 836'-03"
N 824'-03"
N 812'-03"
N 800'-03"
W636'-00"
W 626'-00"
W612'-00"
W600'-00"
AL-1
N 819'-09"
W645'-06"
AL-2
W 588'-06"
AL-3
N842'-09"
W 588'-06"
P-101
W630'-00"
P-102
W618'-00"
N 818'-03"
P 103 SUMP PUMP
W 588-06"
W 640'-03"
N 809'-00"
STABALIZER COLUMN
W 643'-09"
W 631'-09"
N 795'-05"
MCC BUILDING
W623'-09"
N 789'-05"
W629'-09"
N 783'-05"

| | | |
|---|---|---|
| ELECTRICAL DRAFTING | | |
| KEY PLAN | | |
| 12345 | | |
| 1/4"= 1'-0" | SK-E-100 | A |

FIGURE 10.3.

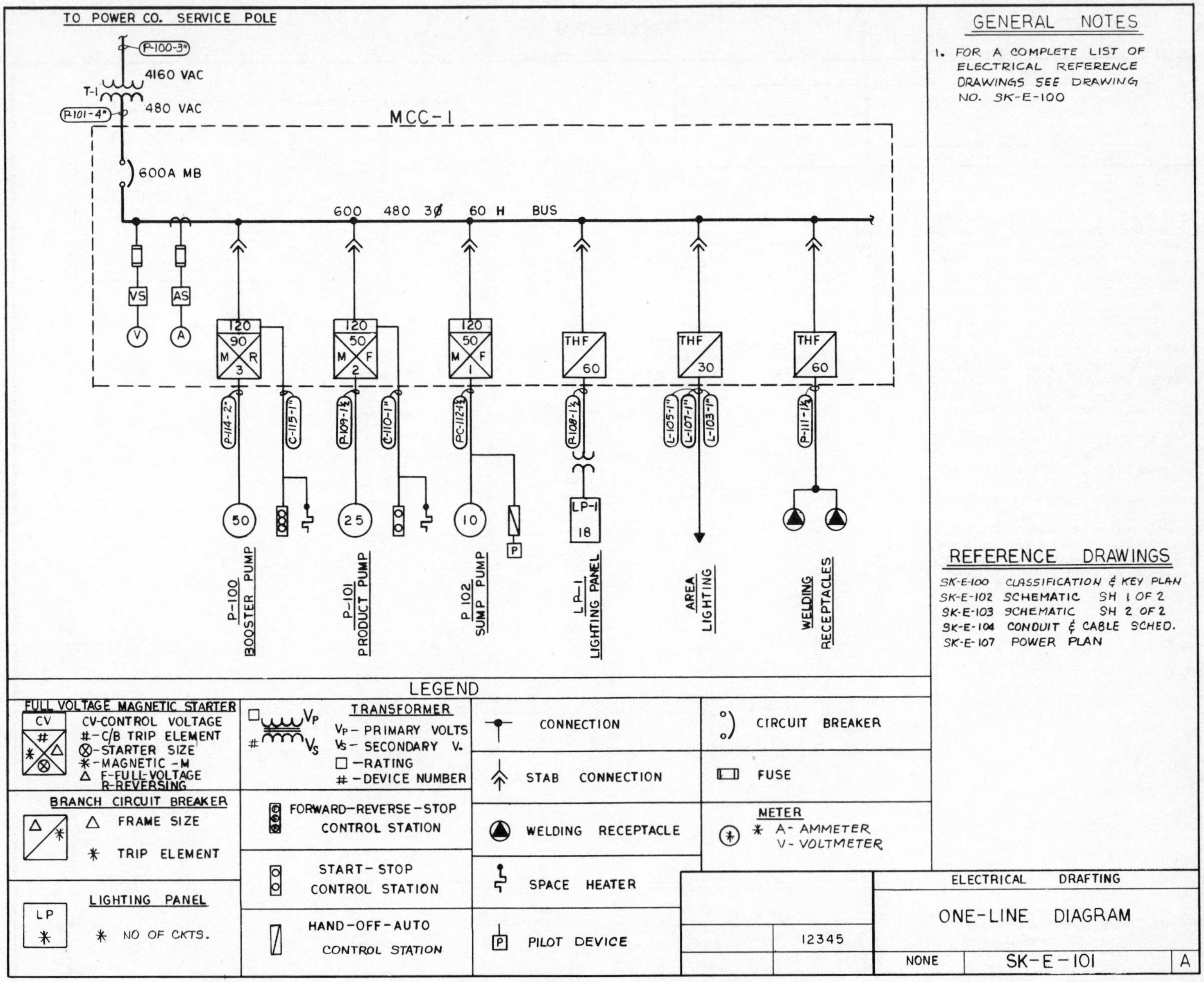

TO POWER CO. SERVICE POLE

P-100-3"

T-1

4160 VAC

480 VAC

P-101-4"

MCC-1

600A MB

600 480 3Ø 60 H BUS

VS

AS

V

A

| 120 / 90 / M R / 3 | 120 / 50 / M F / 2 | 120 / 50 / M F / 1 | THF / 60 | THF / 30 | THF / 60 |
|---|---|---|---|---|---|
| P-114-2" | P-109-1½" | PC-112-1½" | P-108-1½" | L-105-1" | P-111-1½" |
| C-115-1" | C-110-1" | | | L-107-1" | |
| | | | | L-103-1" | |
| 50 | 25 | 10 | LP-1 / 18 | | |
| P-100 BOOSTER PUMP | P-101 PRODUCT PUMP | P 102 SUMP PUMP | LP-1 LIGHTING PANEL | AREA LIGHTING | WELDING RECEPTACLES |

P

## GENERAL NOTES

1. FOR A COMPLETE LIST OF ELECTRICAL REFERENCE DRAWINGS SEE DRAWING NO. SK-E-100

## REFERENCE DRAWINGS

SK-E-100 CLASSIFICATION & KEY PLAN
SK-E-102 SCHEMATIC SH 1 OF 2
SK-E-103 SCHEMATIC SH 2 OF 2
SK-E-104 CONDUIT & CABLE SCHED.
SK-E-107 POWER PLAN

## LEGEND

FULL VOLTAGE MAGNETIC STARTER
CV
CV-CONTROL VOLTAGE
#-C/B TRIP ELEMENT
⊗-STARTER SIZE
*-MAGNETIC -M
△ F-FULL-VOLTAGE
R-REVERSING

TRANSFORMER
$V_P$ - PRIMARY VOLTS
$V_S$ - SECONDARY V.
□ -RATING
# -DEVICE NUMBER

CONNECTION

CIRCUIT BREAKER

BRANCH CIRCUIT BREAKER
△ FRAME SIZE
* TRIP ELEMENT

FORWARD-REVERSE-STOP CONTROL STATION

STAB CONNECTION

FUSE

WELDING RECEPTACLE

METER
* A- AMMETER
V- VOLTMETER

LIGHTING PANEL
LP
* NO OF CKTS.

START-STOP CONTROL STATION

SPACE HEATER

HAND-OFF-AUTO CONTROL STATION

PILOT DEVICE

12345

ELECTRICAL DRAFTING

ONE-LINE DIAGRAM

NONE | SK-E-101 | A

**FIGURE 10.4.** Method A

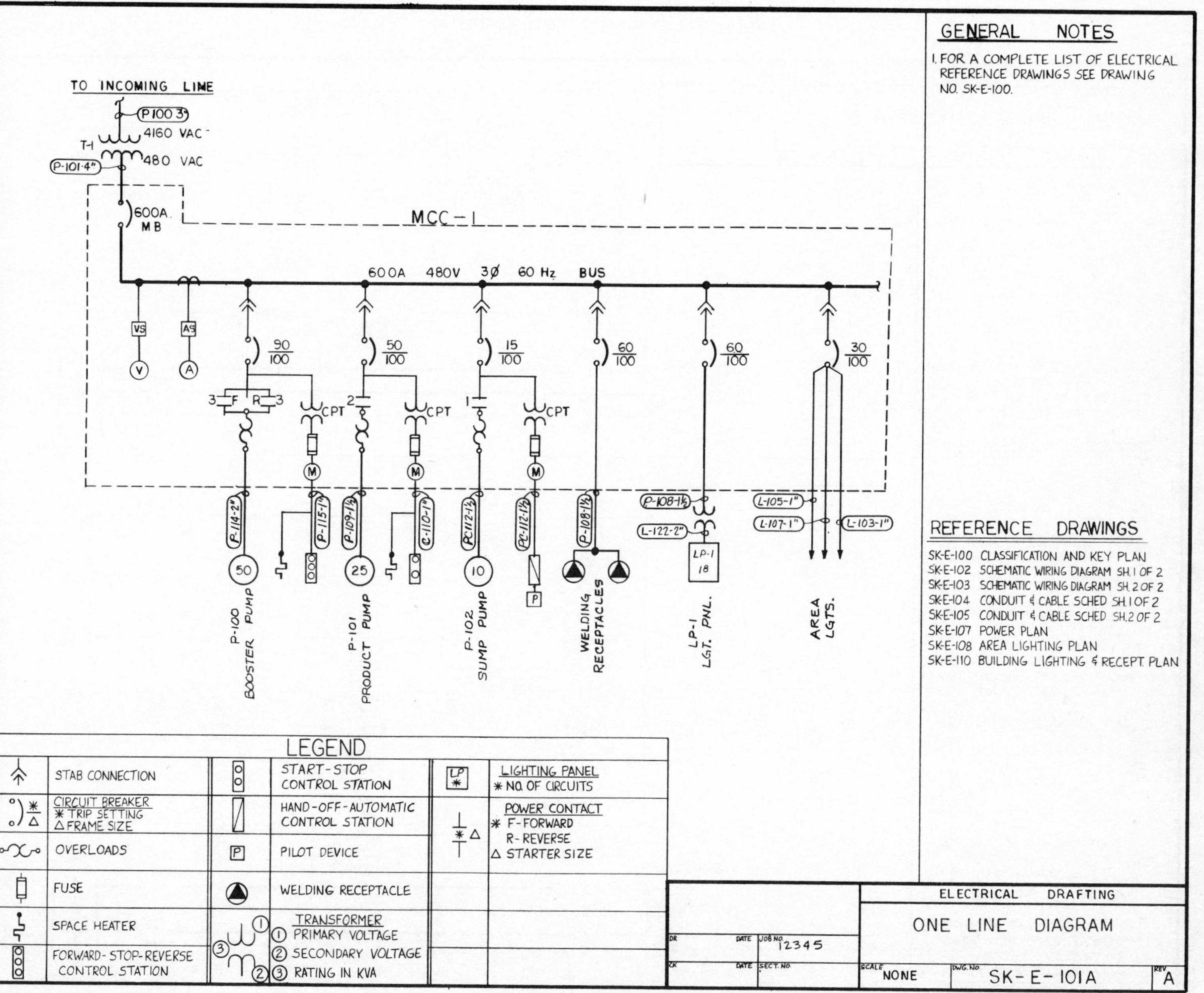


FIGURE 10.5. Method B

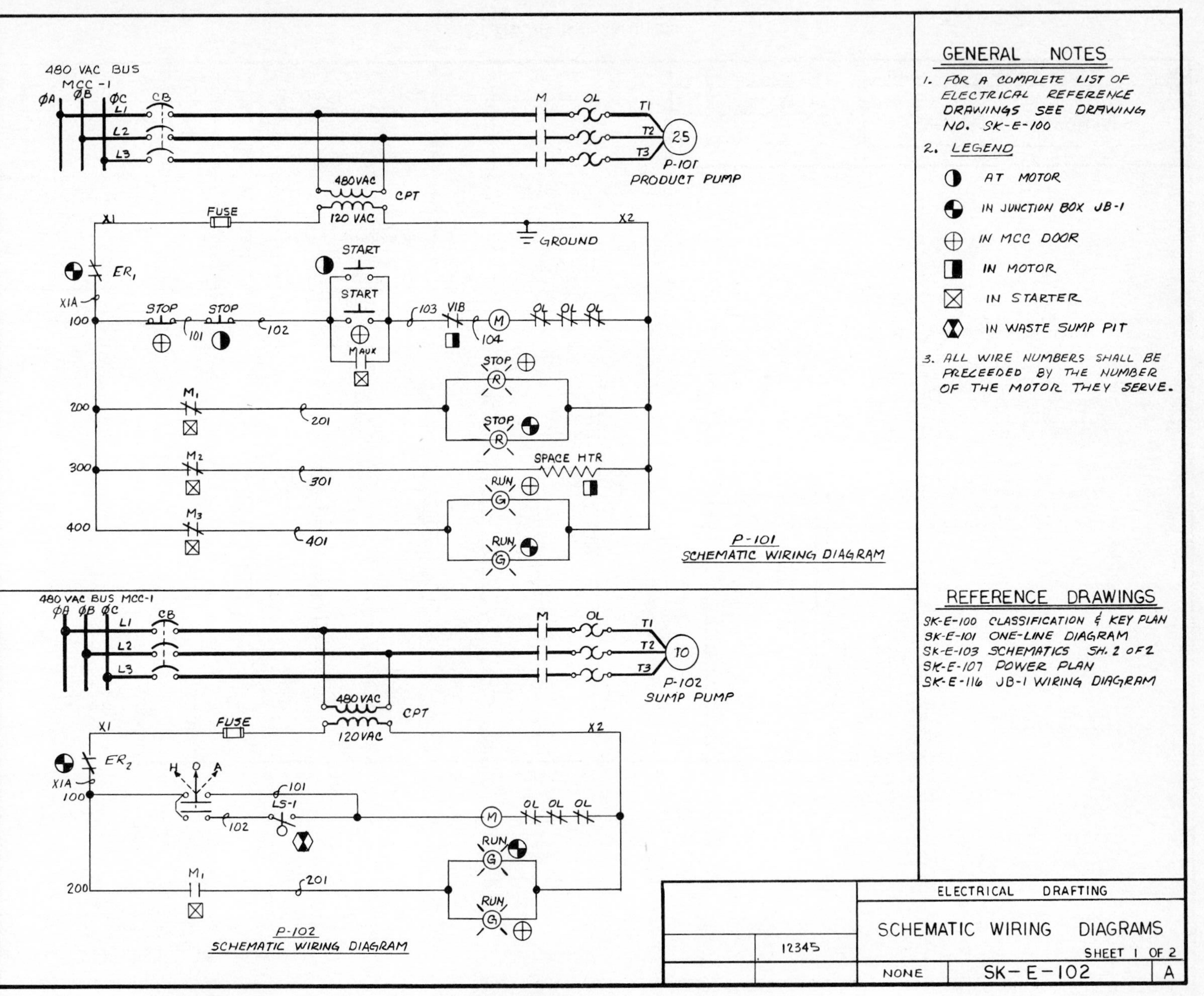


**FIGURE 10.6.** Schematic Wiring Diagrams

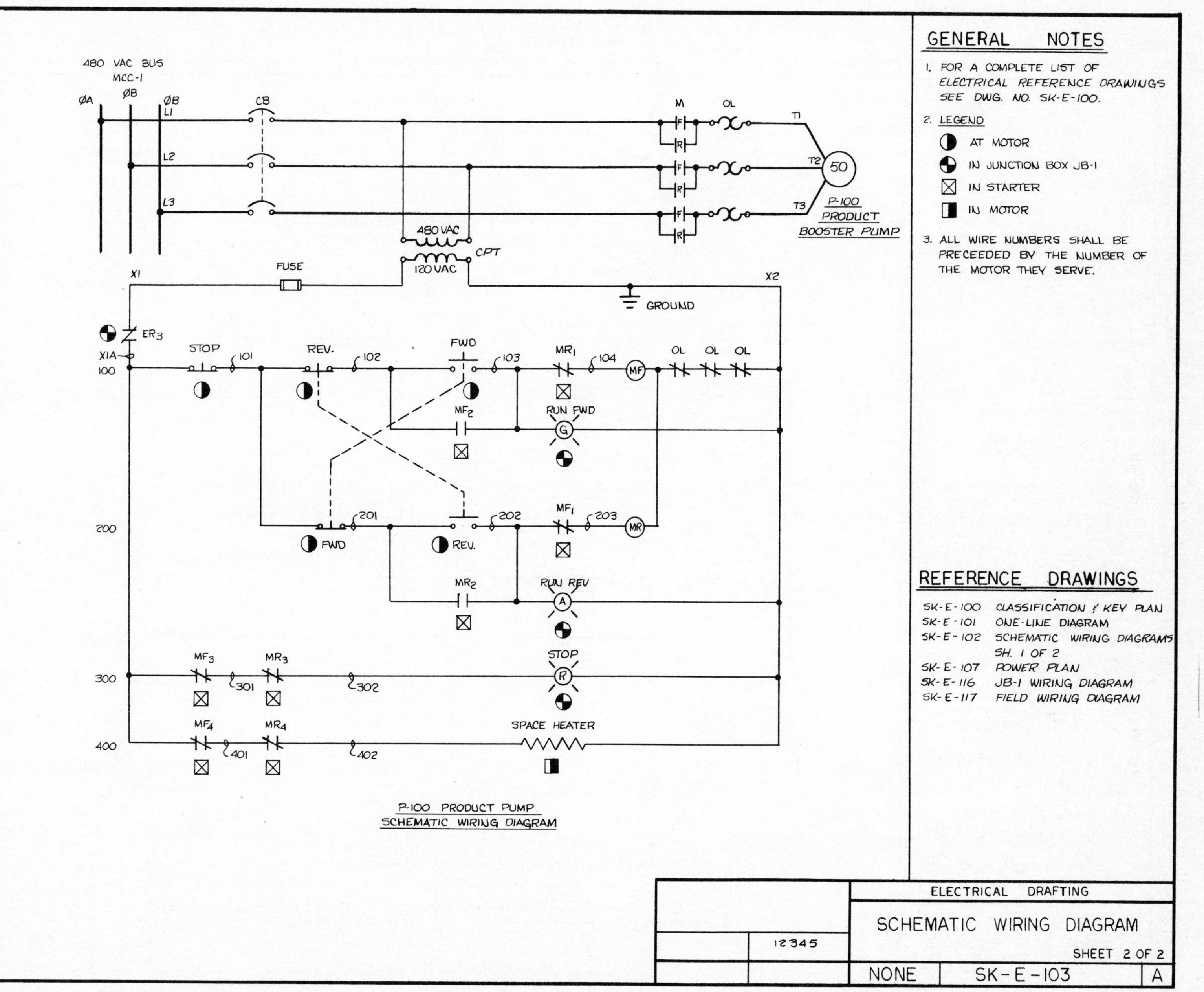


**FIGURE 10.7.** Method A

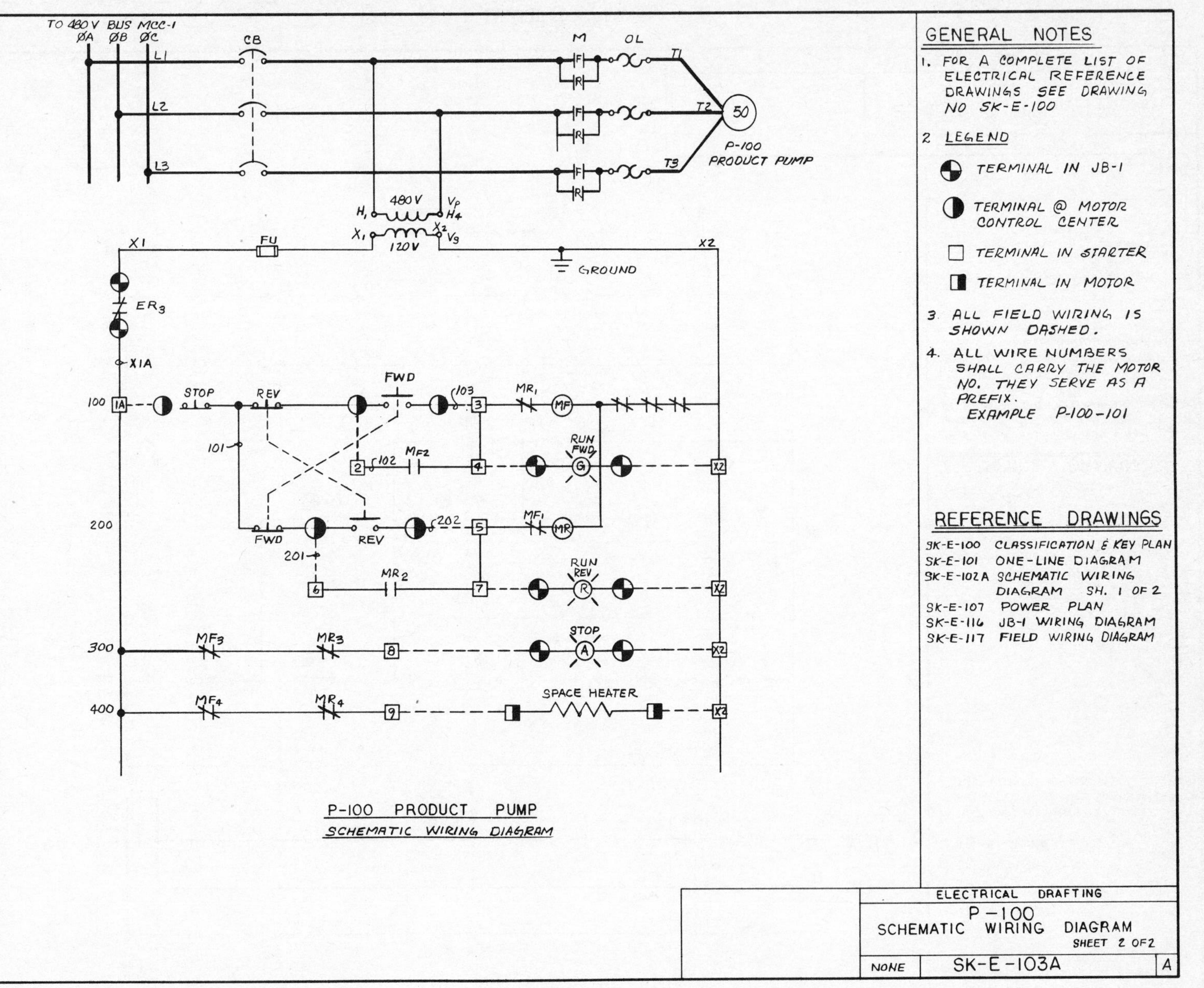


**FIGURE 10.8.** Method B

## CONDUIT AND CABLE SCHEDULE

| CONDUIT NO. | SIZE | TYPE | FROM | TO | VOLTAGE | POWER NO. | POWER SIZE | CONTROL NO | CONTROL SIZE | LENGTH | DEVICE | DWG NO | REMARKS |
|---|---|---|---|---|---|---|---|---|---|---|---|---|---|
| P-100 | 3" | PCV | POWER COMPANY SERVICE POLE | PRIMARY-XFMR T1 | 4160V | 3 | #4(5KU) | — | — | 50' | T-1 | SK-E 107 | INCOMING POWER |
| P-101 | 4" | PCV | SECONDARY SIDE XFMR T1 | MAIN BREAKER MCC-1 CUBICLE | 480V | 6 | 350 MCM | — | — | 10' | MCC-1 | SK-E 107 | MCC-1 SERVICE |
| P-102 | 2" | RGS | SECONDARY XFMR T2 | LGT. PANEL LP-1 | 120/208 | 3 | #2/0 | — | — | 2' | XFMR T-1 | SK-E 108 | LP-1 SERVICE |
| L-103 | 1" | PCV | MCC-1 CUBICLE 2C-1 | AREA LGT. POLE AL-3 | 480V | 3/C | #10 | — | — | 100' | AREA LGT. | SK-E 108 | VIA CABLE TRAY |
| L-104 | 1" | PCV | LGT. PANEL LP-1 | STABILIZER COL. LIGHTING | 120V | 3 | #12 | — | — | 20' | LGT. | SK-E 108 | UNDERGROUND |
| L-105 | 1" | PVC | MCC-1 CUBICLE 2C-1 | AREA LGT. POLE AL-1 | 480V | 3/C | #10 | — | — | 50' | AREA LGT. | SK-E 108 | |
| L-106 | 1" | PVC | LGT. PANEL LP-1 | RACK LIGHTING CONTRACTOR COL. A-1 | 120V | 3/C | #10 | — | — | 120' | AREA LGT-CONT | SK-E 108 | VIA CABLE TRAY |
| L-107 | 1" | PVC | MCC-1 CUBICLE 2C-1 | AREA LGT. POLE AL-2 | 480V | 3/C | #10 | — | — | 60' | AREA LGT. | SK-E 108 | |
| P-108 | 1½" | PVC | MCC-1 CUBICLE 2B-1 | PRIMARY XFMR T-2 CONTROL BLDG. | 480V | 3 | #6 | — | — | 12' | T-2 | SK-E 107 | |
| P-109 | 1½" | PVC | MCC-1 CUBICLE 2E | MOTOR CONNECTION BOX-P-101 | 480V | 3/C | #8 | — | — | 80' | P-101 | SK-E 107 | VIA CABLE TRAY |
| C-110 | 1" | PVC | MCC-1 CUBICLE 2E | CONTROL STATION @-P-101 | 120V | — | — | 6/C | #12 | 80' | P-101 | SK-E 107 | VIA CABLE TRAY |
| P-111 | 1½" | PVC | MCC-1 CUBICLE 2C-2 | WELDING RECEPT. COL. C-3 | 480V | 3/C | #6 | — | — | 65' | WELD RECEPT | SK-E 107 | VIA CABLE TRAY |
| PC-112 | 1" | PVC | MCC-1 CUBICLE 2B-1 | MOTOR CONNECTION BOX-P-102 | 480/120 | 3/C | #12 | 5/C | #12 | 80' | P-102 | SK-E 107 | VIA CABLE TRAY |
| SP-113 | 1½" | PVC | MCC-1 CUBICLE 2B-2 | STUB-OUT & CAP N. SIDE BLDG. | — | — | — | — | — | 12' | — | SK-E 107 | SPARE |
| P-114 | 2" | PVC | MCC-1 CUBICLE 1B | MOTOR CONNECTION BOX-P-100 | 480V | 3/C | #4 | — | — | 85' | P-100 | SK-E 107 | VIA CABLE TRAY |
| C-115 | 1" | PVC | MCC-1 CUBICLE 1B | CONTROL STATION @-P-100 | 120V | — | — | 6/C | #12 | 85' | P-100 | SK-E 107 | VIA CABLE TRAY |
| SP-116 | 1½" | PVC | MCC-1 CUBICLE 1 | STUB-OUT & CAP N. SIDE BLDG. | — | — | — | — | — | 12' | — | SK-E 107 | SPARE |
| SC-117 | 1" | PVC | MCC-1 CUBICLE 1 | STUB-OUT & CAP N. SIDE BLDG. | — | — | — | — | — | 12' | — | SK-E 107 | SPARE |
| SP-118 | 1½" | PVC | MCC-1 CUBICLE 2 | STUB-OUT & CAP N. SIDE BLDG. | — | — | — | — | — | 12' | — | SK-E 107 | SPARE |
| SC-119 | 1" | PVC | MCC-1 CUBICLE 2 | STUB-OUT & CAP N. SIDE BLDG. | — | — | — | — | — | 12' | — | SK-E 107 | SPARE |

LEGEND:

P - POWER
C - CONTROL
L - LIGHTING
I - INSTRUMENTATION
S - SPARE
G - GROUNDING

### GENERAL NOTES

1. FOR A COMPLETE LIST OF ELECTRICAL REFERENCE DRAWINGS SEE DWG. NO. SK-E-100.

### REFERENCE DRAWINGS

SK-E-100 CLASSIFICATION & KEY PLAN
SK-E-101 ONE-LINE DIAGRAM
SK-E-107 POWER PLAN
SK-E-108 AREA PLAN
SK-E-111 CONDUIT STUB-UP PLAN MCC BLDG
SK-E-115 CABLE TRAY DETAILS
SK-E-116 JB-1 WIRING DIAGRAM
SK-E-117 FIELD WIRING DIAGRAM
SK-E-118 MISC. DETAILS & ELEVATIONS SH. 1 OF 2

ELECTRICAL DRAFTING

CONDUIT AND CABLE SCHEDULE

12345

SHEET 1 OF 2

NONE | SK-E-104 | A

FIGURE 10.9.

## CONDUIT AND CABLE SCHEDULE

| CONDUIT NO. | SIZE | TYPE | FROM | TO | VOLTAGE | POWER NO. | POWER SIZE | CONTROL NO. | CONTROL SIZE | LENGTH | DEVICE | DWG. NO. | REMARKS |
|---|---|---|---|---|---|---|---|---|---|---|---|---|---|
| P-120 | 1" | PVC | LGT. PANEL LP-1 | JB-1 POWER | — | 3 | #12 | — | — | — | JB-1 | SK-E 107 | JB-1 POWER VIA CABLE TRAY |
| S-121 | 1" | PVC | LGT. PANEL LP-1 | STUB-OUT & CAP N. SIDE OF BLDG. | — | — | — | — | — | 12' | — | SK-E 107 | SPARE |
| | | | | | | | | | | | | | |
| | | | | | | | | | | | | | |
| | | | | | | | | | | | | | |
| | | | | | | | | | | | | | |
| | | | | | | | | | | | | | |
| | | | | | | | | | | | | | |
| | | | | | | | | | | | | | |
| | | | | | | | | | | | | | |
| | | | | | | | | | | | | | |
| | | | | | | | | | | | | | |
| | | | | | | | | | | | | | |
| | | | | | | | | | | | | | |
| | | | | | | | | | | | | | |
| | | | | | | | | | | | | | |
| | | | | | | | | | | | | | |
| | | | | | | | | | | | | | |
| | | | | | | | | | | | | | |
| | | | | | | | | | | | | | |
| | | | | | | | | | | | | | |
| | | | | | | | | | | | | | |
| | | | | | | | | | | | | | |
| | | | | | | | | | | | | | |

LEGEND:

P - POWER
C - CONTROL
L - LIGHTING
I - INSTRUMENTATION
S - SPARE
G - GROUNDING

### GENERAL NOTES

1. FOR A COMPLETE LIST OF ALL ELECTRICAL REFERENCE DRAWINGS SEE DWG. NO. SK-E-100

### REFERENCE DRAWINGS

SK-E-100 CLASSIFICATION & KEY PLAN
SK-E-104 CONDUIT & CABLE SCHEDULE SH 1 OF 2
SK-E-111 CONDUIT STUB-UP PLAN MCC BLDG.

ELECTRICAL DRAFTING

CONDUIT AND CABLE SCHDULE

12345

SHEET 2 OF 2

NONE | SK-E-105 | A

FIGURE 10.10.

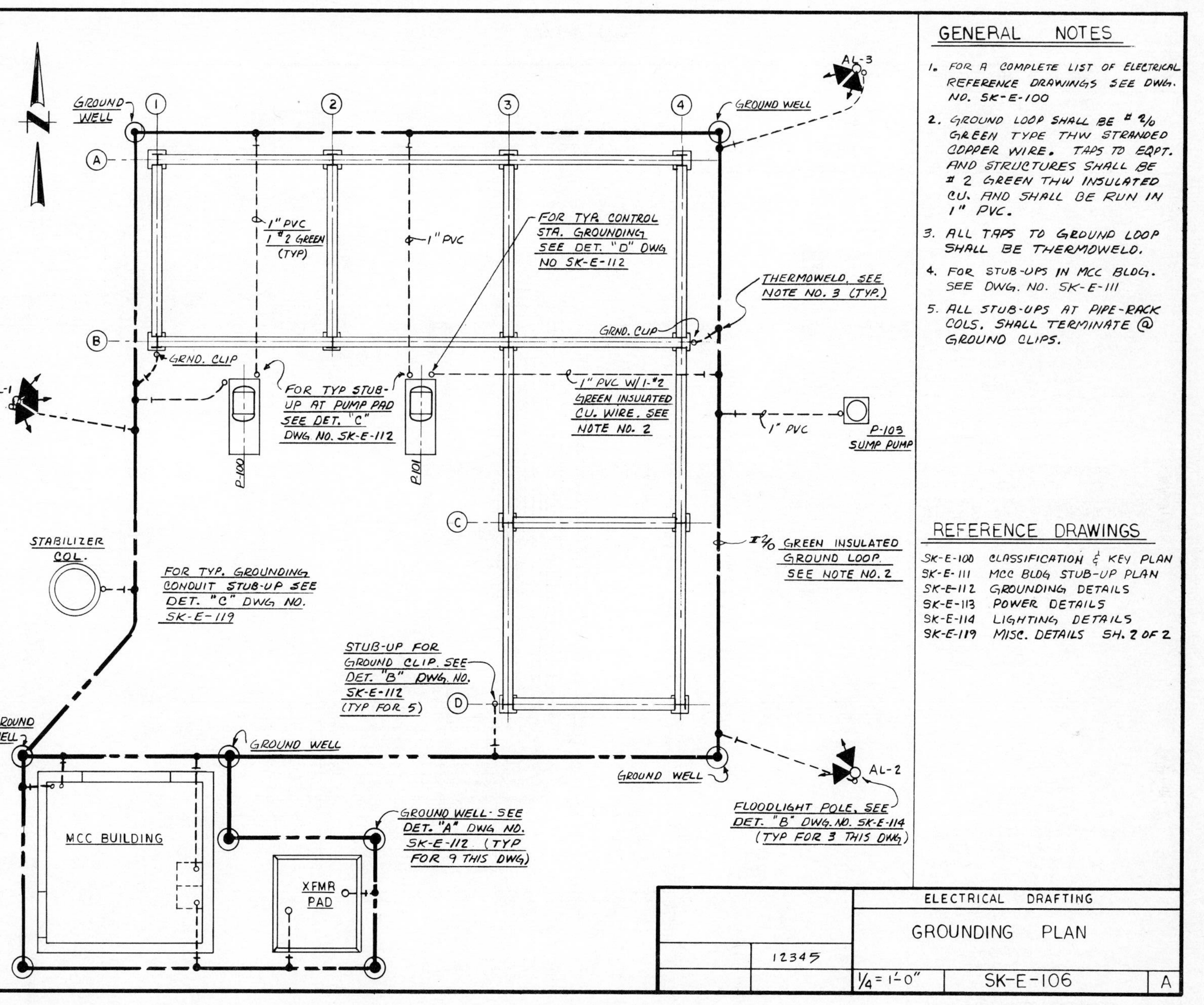
GENERAL NOTES
1. FOR A COMPLETE LIST OF ELECTRICAL REFERENCE DRAWINGS SEE DWG. NO. SK-E-100
2. GROUND LOOP SHALL BE # 2/0 GREEN TYPE THW STRANDED COPPER WIRE. TAPS TO EQPT. AND STRUCTURES SHALL BE # 2 GREEN THW INSULATED CU. AND SHALL BE RUN IN 1" PVC.
3. ALL TAPS TO GROUND LOOP SHALL BE THERMOWELD.
4. FOR STUB-UPS IN MCC BLDG. SEE DWG. NO. SK-E-111
5. ALL STUB-UPS AT PIPE-RACK COLS. SHALL TERMINATE @ GROUND CLIPS.
REFERENCE DRAWINGS
SK-E-100 CLASSIFICATION & KEY PLAN
SK-E-111 MCC BLDG STUB-UP PLAN
SK-E-112 GROUNDING DETAILS
SK-E-113 POWER DETAILS
SK-E-114 LIGHTING DETAILS
SK-E-119 MISC. DETAILS SH. 2 OF 2
GROUND WELL
AL-1
AL-2
AL-3
1
2
3
4
A
B
C
D
1" PVC
1 #2 GREEN
(TYP)
1" PVC
FOR TYP. CONTROL STA. GROUNDING SEE DET. "D" DWG NO SK-E-112
THERMOWELD, SEE NOTE NO. 3 (TYP.)
GRND. CLIP
GRND. CLIP
FOR TYP STUB-UP AT PUMP PAD SEE DET. "C" DWG NO. SK-E-112
1" PVC W/1-#2 GREEN INSULATED CU. WIRE. SEE NOTE NO. 2
1" PVC
P-103
SUMP PUMP
P-100
P-101
STABILIZER COL.
FOR TYP. GROUNDING CONDUIT STUB-UP SEE DET. "C" DWG NO. SK-E-119
#2/0 GREEN INSULATED GROUND LOOP. SEE NOTE NO. 2
STUB-UP FOR GROUND CLIP. SEE DET. "B" DWG. NO. SK-E-112 (TYP FOR 5)
MCC BUILDING
XFMR PAD
GROUND WELL - SEE DET. "A" DWG NO. SK-E-112 (TYP FOR 9 THIS DWG)
FLOODLIGHT POLE. SEE DET. "B" DWG. NO. SK-E-114 (TYP FOR 3 THIS DWG)
ELECTRICAL DRAFTING
GROUNDING PLAN
12345
1/4 = 1'-0"
SK-E-106
A

FIGURE 10.11.

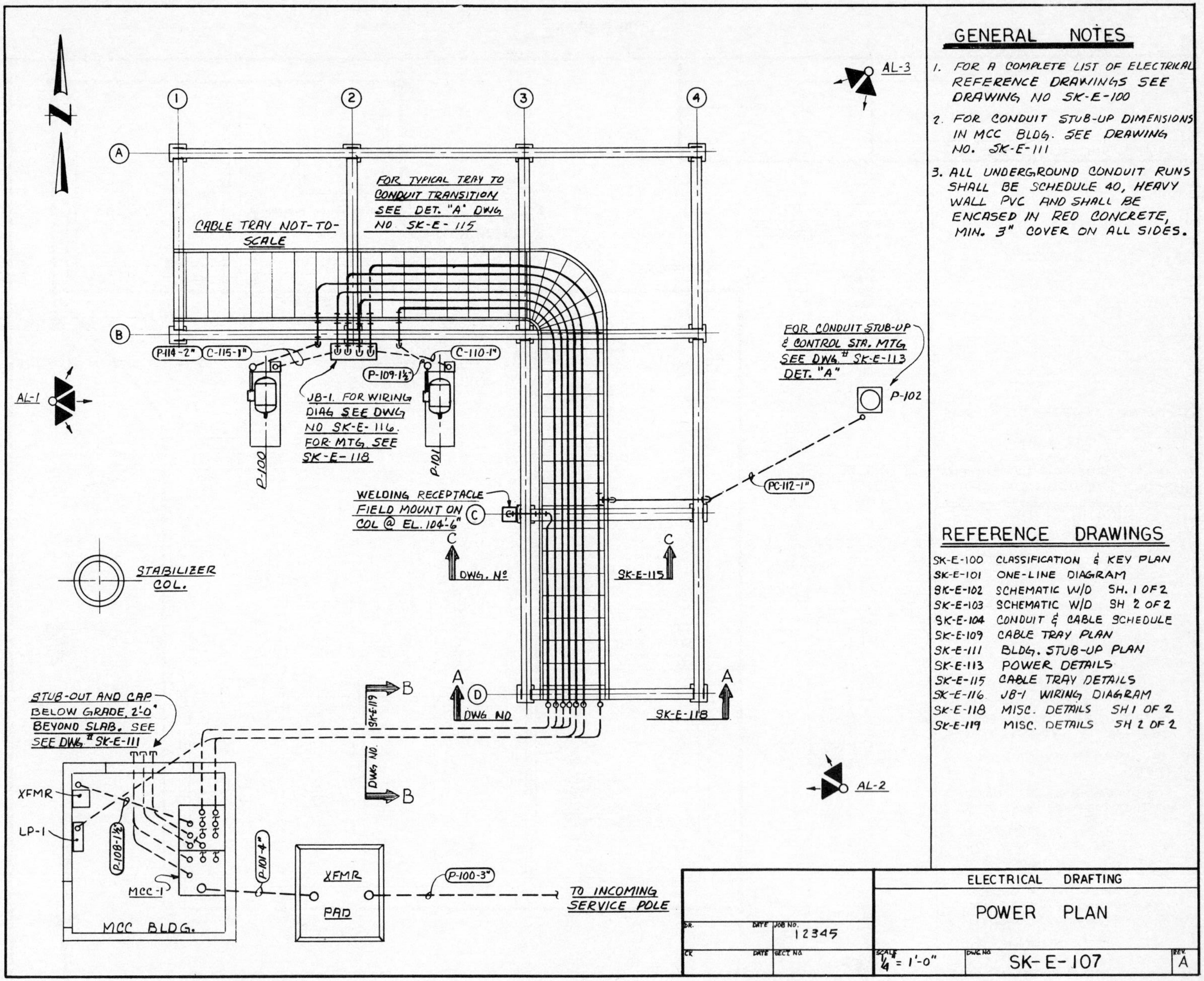

## GENERAL NOTES

1. FOR A COMPLETE LIST OF ELECTRICAL REFERENCE DRAWINGS SEE DRAWING NO SK-E-100
2. FOR CONDUIT STUB-UP DIMENSIONS IN MCC BLDG. SEE DRAWING NO. SK-E-111
3. ALL UNDERGROUND CONDUIT RUNS SHALL BE SCHEDULE 40, HEAVY WALL PVC AND SHALL BE ENCASED IN RED CONCRETE, MIN. 3" COVER ON ALL SIDES.

## REFERENCE DRAWINGS

| | |
|---|---|
| SK-E-100 | CLASSIFICATION & KEY PLAN |
| SK-E-101 | ONE-LINE DIAGRAM |
| SK-E-102 | SCHEMATIC W/D SH. 1 OF 2 |
| SK-E-103 | SCHEMATIC W/D SH 2 OF 2 |
| SK-E-104 | CONDUIT & CABLE SCHEDULE |
| SK-E-109 | CABLE TRAY PLAN |
| SK-E-111 | BLDG. STUB-UP PLAN |
| SK-E-113 | POWER DETAILS |
| SK-E-115 | CABLE TRAY DETAILS |
| SK-E-116 | JB-1 WIRING DIAGRAM |
| SK-E-118 | MISC. DETAILS SH 1 OF 2 |
| SK-E-119 | MISC. DETAILS SH 2 OF 2 |

N
1
2
3
4
A
B
AL-1
AL-2
AL-3
CABLE TRAY NOT-TO-SCALE
FOR TYPICAL TRAY TO CONDUIT TRANSITION SEE DET. "A" DWG NO SK-E-115
P-114-2"
C-115-1"
C-110-1"
P-109-1½"
JB-1. FOR WIRING DIAG SEE DWG NO SK-E-116. FOR MTG. SEE SK-E-118
P-100
P-101
FOR CONDUIT STUB-UP & CONTROL STA. MTG. SEE DWG # SK-E-113 DET. "A"
P-102
PC-112-1"
WELDING RECEPTACLE FIELD MOUNT ON COL @ EL. 104'-6"
C
DWG. No
SK-E-115
STABILIZER COL.
STUB-OUT AND CAP BELOW GRADE, 2'-0" BEYOND SLAB. SEE SEE DWG # SK-E-111
B
SK-E-119
DWG NO.
D
DWG NO
SK-E-118
XFMR
LP-1
P-108-1½"
MCC-1
MCC BLDG.
P-101-4"
XFMR
PAD
P-100-3"
TO INCOMING SERVICE POLE

| | | | |
|---|---|---|---|
| DR. | DATE | JOB NO. 12345 | ELECTRICAL DRAFTING |
| | | | POWER PLAN |
| CK | DATE | SECT. NO. | SCALE ¼" = 1'-0" · DWG NO SK-E-107 · REV. A |

FIGURE 10.12.

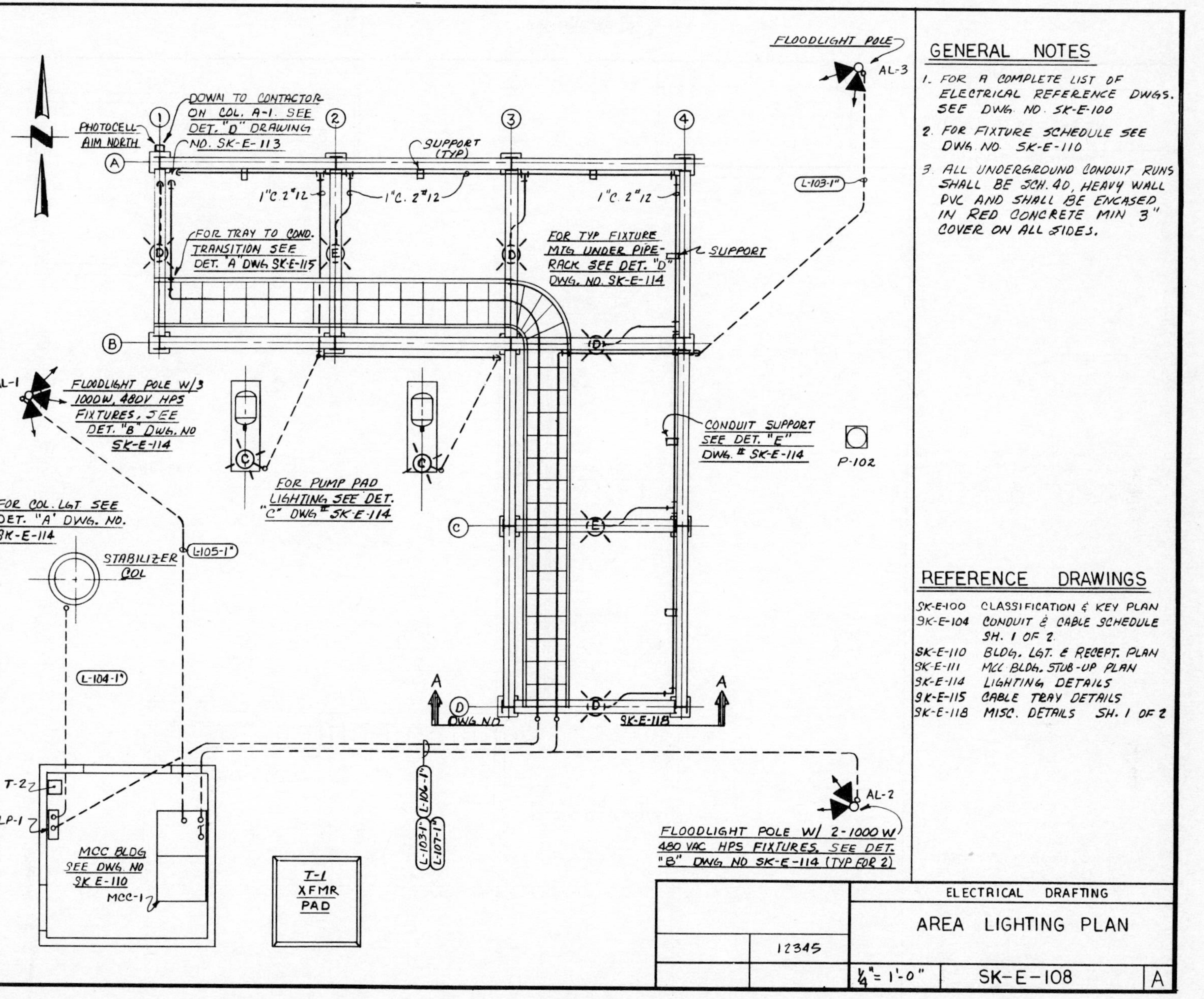
FLOODLIGHT POLE
AL-3
GENERAL NOTES
1. FOR A COMPLETE LIST OF ELECTRICAL REFERENCE DWGS. SEE DWG. NO. SK-E-100
2. FOR FIXTURE SCHEDULE SEE DWG NO. SK-E-110
3. ALL UNDERGROUND CONDUIT RUNS SHALL BE SCH. 40, HEAVY WALL PVC AND SHALL BE ENCASED IN RED CONCRETE MIN 3" COVER ON ALL SIDES.
DOWN TO CONTACTOR ON COL. A-1. SEE DET. "D" DRAWING NO. SK-E-113
PHOTOCELL AIM NORTH
SUPPORT (TYP)
1"C. 2"12
L-103-1"
FOR TRAY TO COND. TRANSITION SEE DET. "A" DWG SK-E-115
FOR TYP FIXTURE MTG UNDER PIPE-RACK SEE DET. "D" DWG. NO. SK-E-114
SUPPORT
AL-1
FLOODLIGHT POLE W/3 1000W, 480V HPS FIXTURES, SEE DET. "B" DWG. NO SK-E-114
CONDUIT SUPPORT SEE DET. "E" DWG. # SK-E-114
P-102
FOR PUMP PAD LIGHTING SEE DET. "C" DWG # SK-E-114
FOR COL. LGT SEE DET. "A" DWG. NO. SK-E-114
STABILIZER COL
L-105-1"
REFERENCE DRAWINGS
SK-E-100 CLASSIFICATION & KEY PLAN
SK-E-104 CONDUIT & CABLE SCHEDULE SH. 1 OF 2
SK-E-110 BLDG. LGT. & RECEPT. PLAN
SK-E-111 MCC BLDG. STUB-UP PLAN
SK-E-114 LIGHTING DETAILS
SK-E-115 CABLE TRAY DETAILS
SK-E-118 MISC. DETAILS SH. 1 OF 2
L-104-1"
DWG NO
SK-E-118
T-2
LP-1
L-106-1"
L-103-1"
L-107-1"
AL-2
FLOODLIGHT POLE W/ 2-1000 W 480 VAC HPS FIXTURES. SEE DET. "B" DWG NO SK-E-114 (TYP FOR 2)
MCC BLDG SEE DWG NO SK E-110
MCC-1
T-1 XFMR PAD
ELECTRICAL DRAFTING
AREA LIGHTING PLAN
12345
1/4" = 1'-0"
SK-E-108
A

FIGURE 10.13.

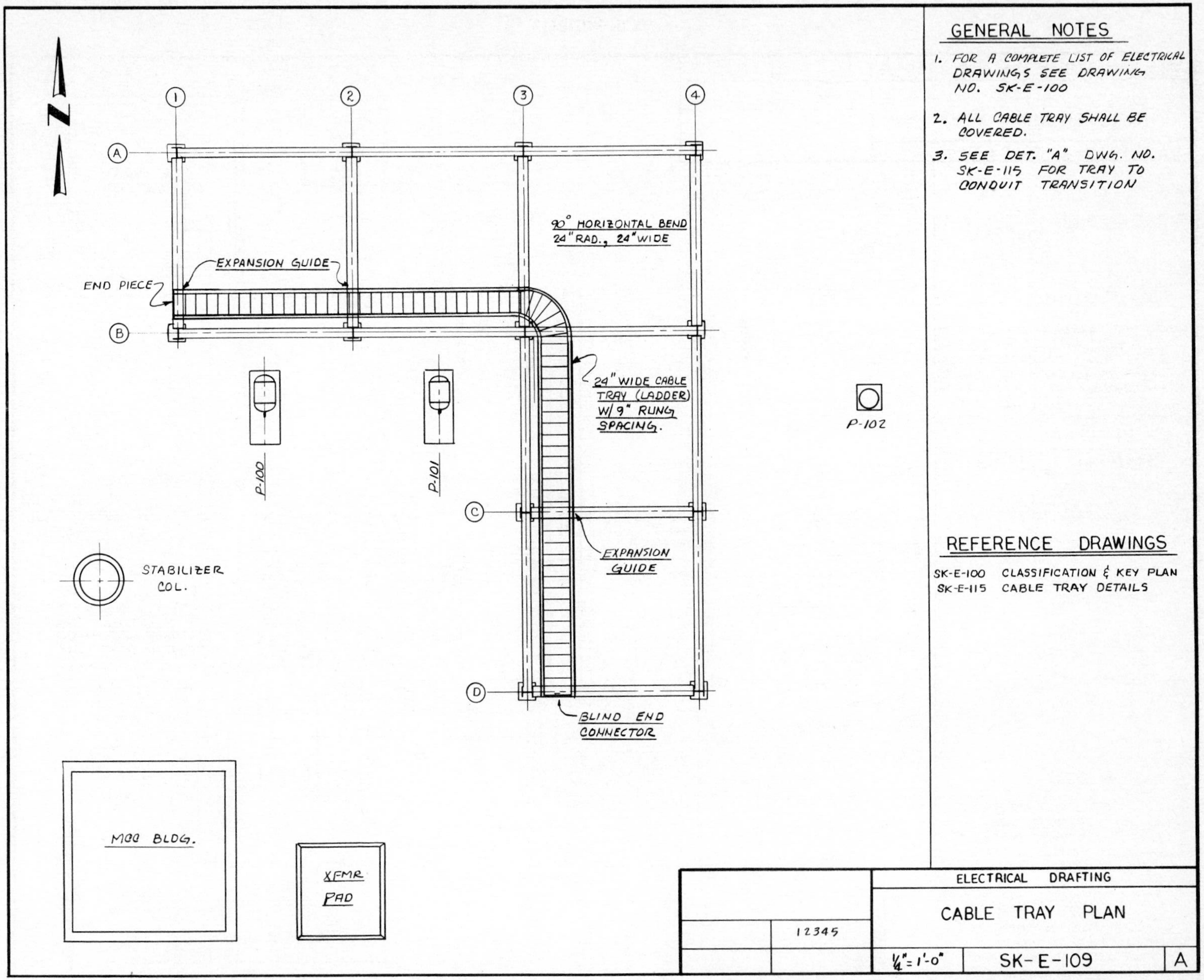

GENERAL NOTES

1. FOR A COMPLETE LIST OF ELECTRICAL DRAWINGS SEE DRAWING NO. SK-E-100
2. ALL CABLE TRAY SHALL BE COVERED.
3. SEE DET. "A" DWG. NO. SK-E-115 FOR TRAY TO CONDUIT TRANSITION

REFERENCE DRAWINGS

SK-E-100 CLASSIFICATION & KEY PLAN
SK-E-115 CABLE TRAY DETAILS

N
1
2
3
4
A
B
C
D
90° HORIZONTAL BEND 24" RAD., 24" WIDE
EXPANSION GUIDE
END PIECE
24" WIDE CABLE TRAY (LADDER) W/ 9" RUNG SPACING.
P-100
P-101
P-102
EXPANSION GUIDE
STABILIZER COL.
BLIND END CONNECTOR
MCC BLDG.
XFMR PAD

| | | ELECTRICAL DRAFTING | |
|---|---|---|---|
| | | CABLE TRAY PLAN | |
| | 12345 | | |
| | | ¼" = 1'-0" SK-E-109 | A |

FIGURE 10.14.

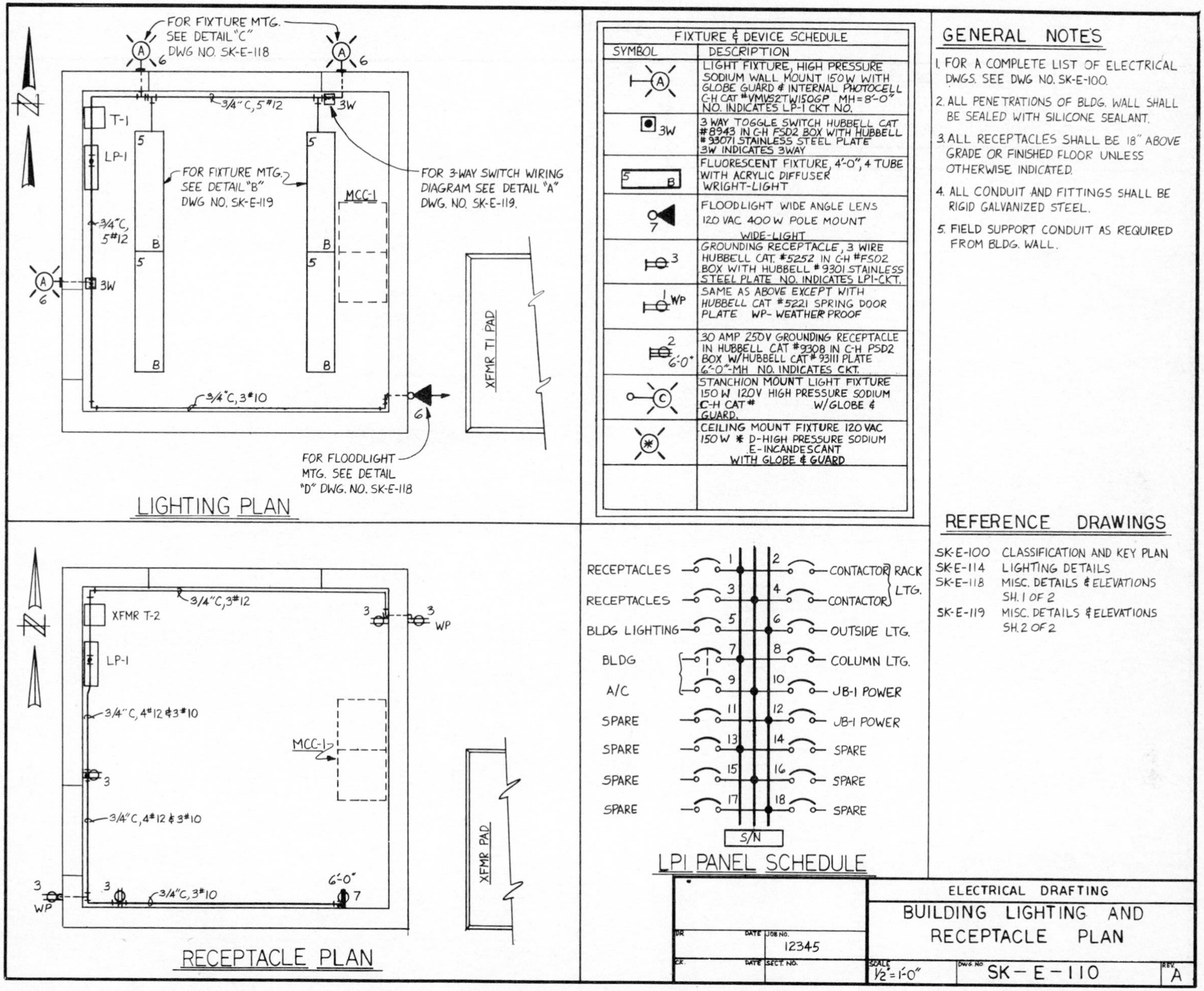

## LIGHTING PLAN

FOR FIXTURE MTG. SEE DETAIL "C" DWG NO. SK-E-118

FOR FIXTURE MTG. SEE DETAIL "B" DWG NO. SK-E-119

FOR 3-WAY SWITCH WIRING DIAGRAM SEE DETAIL "A" DWG. NO. SK-E-119.

FOR FLOODLIGHT MTG. SEE DETAIL "D" DWG. NO. SK-E-118

T-1
LP-1
MCC-1
3/4" C, 5#12
3/4" C, 3#10
3W
XFMR T1 PAD

## RECEPTACLE PLAN

XFMR T-2
LP-1
MCC-1
3/4" C, 3#12
3/4" C, 4#12 & 3#10
3/4" C, 3#10
6'-0"
WP
XFMR PAD

## FIXTURE & DEVICE SCHEDULE

| SYMBOL | DESCRIPTION |
|---|---|
| (A) | LIGHT FIXTURE, HIGH PRESSURE SODIUM WALL MOUNT 150W WITH GLOBE GUARD & INTERNAL PHOTOCELL C-H CAT #VMVS2TW150GP MH=8'-0" NO. INDICATES LP-1 CKT NO. |
| 3W | 3 WAY TOGGLE SWITCH HUBBELL CAT #8943 IN C-H FSD2 BOX WITH HUBBELL #93071 STAINLESS STEEL PLATE 3W INDICATES 3WAY |
| 5 B | FLUORESCENT FIXTURE, 4'-0", 4 TUBE WITH ACRYLIC DIFFUSER WRIGHT-LIGHT |
| 7 | FLOODLIGHT WIDE ANGLE LENS 120 VAC 400 W POLE MOUNT WIDE-LIGHT |
| 3 | GROUNDING RECEPTACLE, 3 WIRE HUBBELL CAT. #5252 IN C-H #FS02 BOX WITH HUBBELL #9301 STAINLESS STEEL PLATE NO. INDICATES LP1-CKT. |
| WP | SAME AS ABOVE EXCEPT WITH HUBBELL CAT #5221 SPRING DOOR PLATE WP- WEATHER PROOF |
| 2 6'-0" | 30 AMP 250V GROUNDING RECEPTACLE IN HUBBELL CAT #9308 IN C-H PSD2 BOX W/HUBBELL CAT #93111 PLATE 6'-0"-MH NO. INDICATES CKT. |
| (C) | STANCHION MOUNT LIGHT FIXTURE 150 W 120V HIGH PRESSURE SODIUM C-H CAT # W/GLOBE & GUARD. |
| (*) | CEILING MOUNT FIXTURE 120 VAC 150 W * D-HIGH PRESSURE SODIUM E-INCANDESCANT WITH GLOBE & GUARD |
| | |

## GENERAL NOTES

1. FOR A COMPLETE LIST OF ELECTRICAL DWGS. SEE DWG NO. SK-E-100.
2. ALL PENETRATIONS OF BLDG. WALL SHALL BE SEALED WITH SILICONE SEALANT.
3. ALL RECEPTACLES SHALL BE 18" ABOVE GRADE OR FINISHED FLOOR UNLESS OTHERWISE INDICATED.
4. ALL CONDUIT AND FITTINGS SHALL BE RIGID GALVANIZED STEEL.
5. FIELD SUPPORT CONDUIT AS REQUIRED FROM BLDG. WALL.

## REFERENCE DRAWINGS

| | |
|---|---|
| SK-E-100 | CLASSIFICATION AND KEY PLAN |
| SK-E-114 | LIGHTING DETAILS |
| SK-E-118 | MISC. DETAILS & ELEVATIONS SH. 1 OF 2 |
| SK-E-119 | MISC. DETAILS & ELEVATIONS SH. 2 OF 2 |

## LP1 PANEL SCHEDULE

| Load | Ckt | Ckt | Load |
|---|---|---|---|
| RECEPTACLES | 1 | 2 | CONTACTOR (RACK LTG.) |
| RECEPTACLES | 3 | 4 | CONTACTOR (RACK LTG.) |
| BLDG LIGHTING | 5 | 6 | OUTSIDE LTG. |
| BLDG A/C | 7 | 8 | COLUMN LTG. |
| BLDG A/C | 9 | 10 | JB-1 POWER |
| SPARE | 11 | 12 | JB-1 POWER |
| SPARE | 13 | 14 | SPARE |
| SPARE | 15 | 16 | SPARE |
| SPARE | 17 | 18 | SPARE |

S/N

ELECTRICAL DRAFTING

BUILDING LIGHTING AND RECEPTACLE PLAN

DR. DATE JOB NO. 12345
CK. DATE SECT. NO.
SCALE 1/2" = 1'-0" DWG. NO. SK - E - 110 REV. A

FIGURE 10.15.

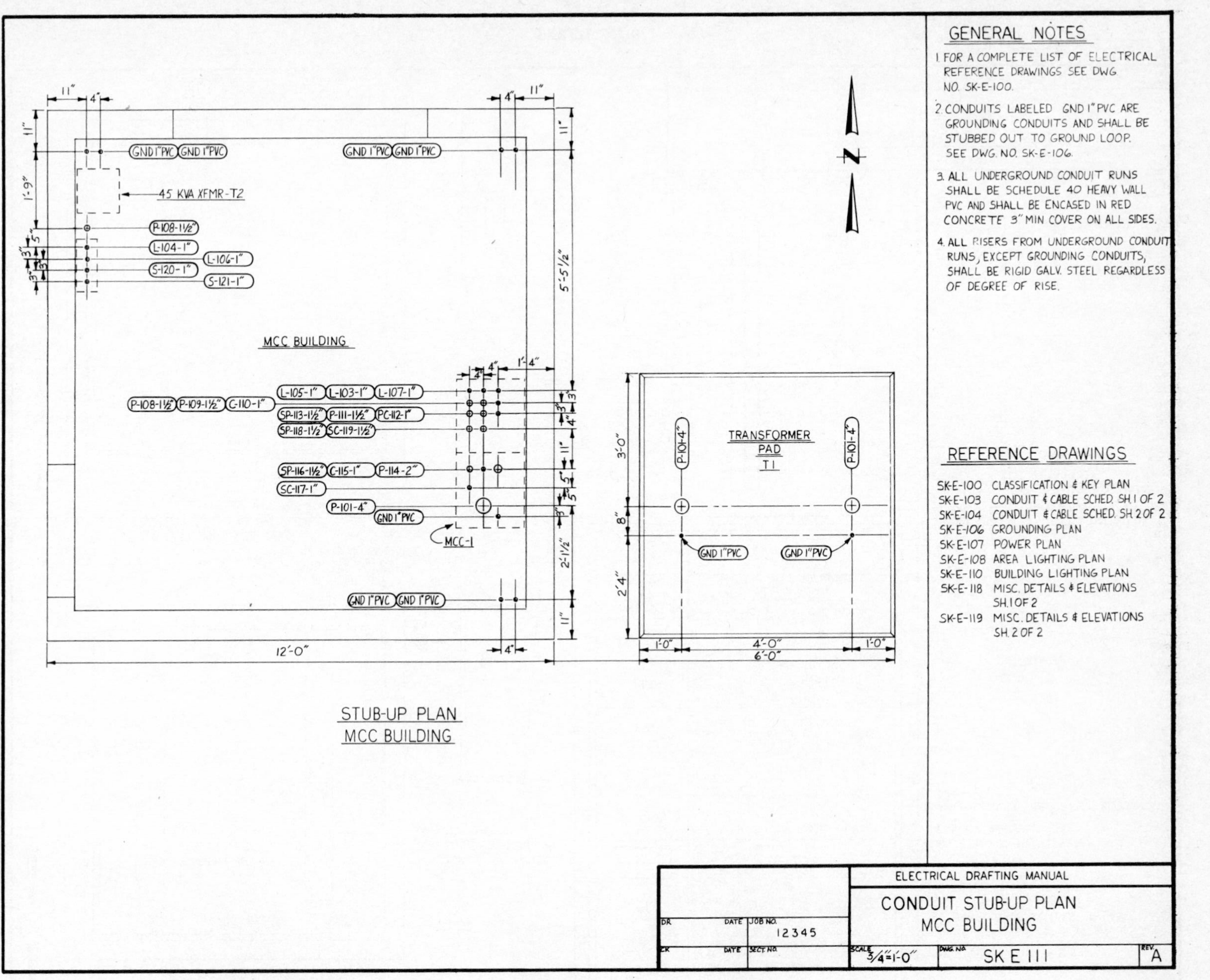
GENERAL NOTES
1. FOR A COMPLETE LIST OF ELECTRICAL REFERENCE DRAWINGS SEE DWG NO. SK-E-100.
2. CONDUITS LABELED GND 1" PVC ARE GROUNDING CONDUITS AND SHALL BE STUBBED OUT TO GROUND LOOP. SEE DWG. NO. SK-E-106.
3. ALL UNDERGROUND CONDUIT RUNS SHALL BE SCHEDULE 40 HEAVY WALL PVC AND SHALL BE ENCASED IN RED CONCRETE 3" MIN COVER ON ALL SIDES.
4. ALL RISERS FROM UNDERGROUND CONDUIT RUNS, EXCEPT GROUNDING CONDUITS, SHALL BE RIGID GALV. STEEL REGARDLESS OF DEGREE OF RISE.
REFERENCE DRAWINGS
SK-E-100 CLASSIFICATION & KEY PLAN
SK-E-103 CONDUIT & CABLE SCHED. SH.1 OF 2
SK-E-104 CONDUIT & CABLE SCHED. SH.2 OF 2
SK-E-106 GROUNDING PLAN
SK-E-107 POWER PLAN
SK-E-108 AREA LIGHTING PLAN
SK-E-110 BUILDING LIGHTING PLAN
SK-E-118 MISC. DETAILS & ELEVATIONS SH.1 OF 2
SK-E-119 MISC. DETAILS & ELEVATIONS SH.2 OF 2
N
MCC BUILDING
45 KVA XFMR-T2
GND 1" PVC
P-108-1½"
L-104-1"
L-106-1"
S-120-1"
S-121-1"
L-105-1"
L-103-1"
L-107-1"
P-108-1½"
P-109-1½"
C-110-1"
SP-113-1½"
P-111-1½"
PC-112-1"
SP-118-1½"
SC-119-1½"
SP-116-1½"
C-115-1"
P-114-2"
SC-117-1"
P-101-4"
MCC-1
TRANSFORMER PAD T1
12'-0"
5'-5½"
2'-1½"
1'-9"
1'-4"
3'-0"
2'-4"
4'-0"
6'-0"
STUB-UP PLAN
MCC BUILDING
ELECTRICAL DRAFTING MANUAL
CONDUIT STUB-UP PLAN
MCC BUILDING
DR
DATE
JOB NO.
12345
CK
SECT NO.
SCALE 3/4"=1'-0"
DWG. NO. SK E 111
REV. A

FIGURE 10.16.

GROUND WELL
DETAIL 'A'
DWG. NO. SK-E-106

COVER
GRADE
4'-0"
8" Φ BY 4'-0" CONCRETE PIPE WITH FLARED END
BURNDY GROUND CLAMP
#2/0 GREEN GROUND LOOP
GROUND ROD

GROUND CLIP
DETAIL 'B'
DWG. NO. SK-E-106

MATERIAL:
14" x 3" x 1/4" STAINLESS PL

FIELD FABRICATE & DRILL AS REQUIRED

2" 6" 2" 3"
3/8" DIA. (TYP.)
TOP VIEW
2" 6" 2" 2"
FRONT VIEW
3" 2"
SIDE VIEW

## GENERAL NOTES

1. FOR A COMPLETE LIST OF ALL ELECTRICAL REFERENCE DRAWINGS SEE DWG. NO. SK-E-100.

TYPICAL MOTOR GROUND
DETAIL 'C'
DWG. NO. SK-E-106

DO NOT TERMINATE GRND. LEAD UNDER MOUNTING BOLT.
FIELD DRILL AND TAP MOTOR BASE FOR BURNDY QB26 STUD POST
PUMP PAD
GRADE
1" PVC W/ 1 - #2 TO GROUND LOOP

CONTROL STATION GROUNDING
DETAIL 'D'
DWG. NO. SK-E-106

2 4"[, 5.4 # CONTROL STATION STANCHION
BURNDY GRND. CLAMP
PUMP PAD
GRADE
FIELD ANCHOR STANCHION TO PAD W/ CONCRETE ANCHORS.
1" PVC W/ 1-#2 TO GROUND LOOP

## REFERENCE DRAWINGS

SK-E-100 CLASSIFICATION & KEY PLAN
SK-E-106 GROUNDING PLAN
SK-E-107 POWER PLAN

| | | ELECTRICAL DRAFTING | | |
|---|---|---|---|---|
| | | GROUNDING DETAILS | | |
| | 12345 | | | |
| | | NONE | SK - E - 112 | A |

**FIGURE 10.17.**

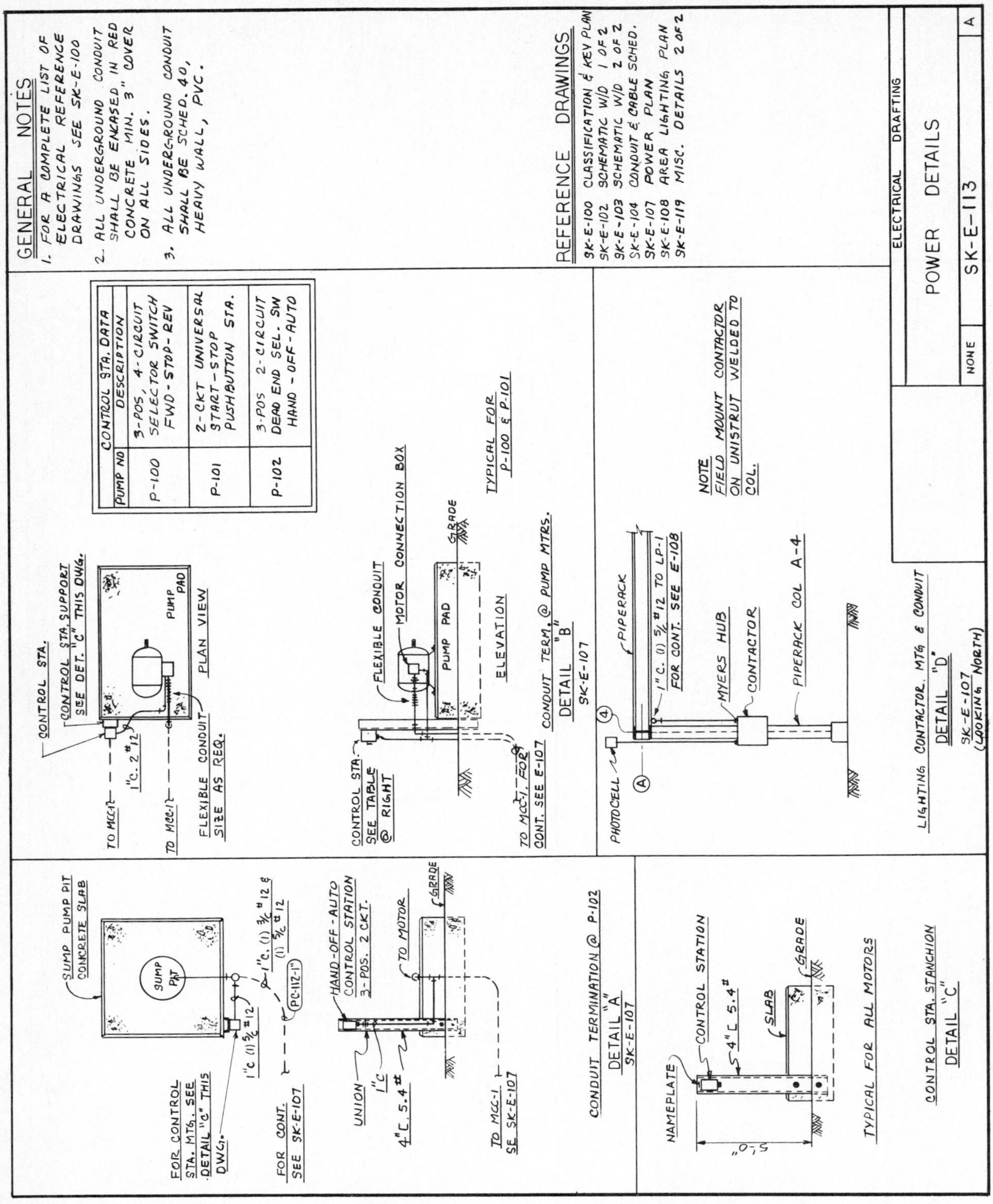

## GENERAL NOTES

1. FOR A COMPLETE LIST OF ELECTRICAL REFERENCE DRAWINGS SEE SK-E-100
2. ALL UNDERGROUND CONDUIT SHALL BE ENCASED IN RED CONCRETE MIN. 3" COVER ON ALL SIDES.
3. ALL UNDERGROUND CONDUIT SHALL BE SCHED. 40, HEAVY WALL, PVC.

## REFERENCE DRAWINGS

SK-E-100 CLASSIFICATION & KEY PLAN
SK-E-102 SCHEMATIC W/D 1 OF 2
SK-E-103 SCHEMATIC W/D 2 OF 2
SK-E-104 CONDUIT & CABLE SCHED.
SK-E-107 POWER PLAN
SK-E-108 AREA LIGHTING PLAN
SK-E-119 MISC. DETAILS 2 OF 2

| CONTROL STA. DATA | |
|---|---|
| PUMP NO | DESCRIPTION |
| P-100 | 3-POS, 4-CIRCUIT SELECTOR SWITCH FWD-STOP-REV |
| P-101 | 2-CKT UNIVERSAL START-STOP PUSHBUTTON STA. |
| P-102 | 3-POS 2-CIRCUIT DEAD END SEL. SW HAND-OFF-AUTO |

SUMP PUMP PIT CONCRETE SLAB
SUMP PIT
FOR CONTROL STA. MTG. SEE DETAIL "C" THIS DWG.
1"C (1) 5/C #12
1"C. (1) 3/C #12 & (1) 5/C #12
FOR CONT. SEE SK-E-107
PC-112-1"
UNION
1"C
4"C. 5.4#
HAND-OFF-AUTO CONTROL STATION 3-POS. 2 CKT.
TO MOTOR
GRADE
TO MCC-1 SE SK-E-107

CONDUIT TERMINATION @ P-102
DETAIL "A"
SK-E-107

CONTROL STA.
CONTROL STA. SUPPORT SEE DET. "C" THIS DWG.
TO MCC-1
1"C. 2 #12
TO MCC-1
PUMP PAD
FLEXIBLE CONDUIT SIZE AS REQ.
PLAN VIEW
CONTROL STA. SEE TABLE @ RIGHT
FLEXIBLE CONDUIT
MOTOR CONNECTION BOX
PUMP PAD
GRADE
ELEVATION
TYPICAL FOR P-100 & P-101
TO MCC-1, FOR CONT. SEE E-107

CONDUIT TERM. @ PUMP MTRS.
DETAIL "B"
SK-E-107

NAMEPLATE
CONTROL STATION
4"C 5.4#
5'-0"
SLAB
GRADE
TYPICAL FOR ALL MOTORS

CONTROL STA. STANCHION
DETAIL "C"

PHOTOCELL
4
A
PIPERACK
1"C. (1) 5/C #12 TO LP-1 FOR CONT. SEE E-108
MYERS HUB
CONTACTOR
PIPERACK COL A-4
NOTE
FIELD MOUNT CONTACTOR ON UNISTRUT WELDED TO COL.

LIGHTING CONTACTOR, MTG & CONDUIT
DETAIL "D"
SK-E-107
(LOOKING NORTH)

ELECTRICAL DRAFTING
POWER DETAILS
NONE | SK-E-113 | A

FIGURE 10.18.

FOR STANCHION SEE DETAIL "C" THIS DRAWING
MH 138'-0"
150 W HPS STANCHION MOUNT FIXTURE C-H CAT # VMS J15 OGP W/GLOBE & GUARD
1 1/4" RGS CONDUIT
STANDOFF - SEE DETAIL "F" THIS DRAWING
C-H REC BELL REDUCER
EL.130'
1" SEALTIGHT W/FITTINGS
1"C, 2#12
FIELD SUPPORT AS REQUIRED
GRADE

STABILIZER COLUMN LIGHTING
DETAIL "A"
SK-E-108

FLOODLIGHT WIDE ANGLE LENS 1000 W 480 VAC W 2" SLIPFITTER
POLE-TOP SUPPORT FOR AL-1 3 FIXTURES
SQUARE, TAPERED HINGED POLE
POLE TOP SUPPORT FOR AL-2 & AL-3 2 FIXTURES
1"PVC, 1#2 GRN GROUND TO GROUND LOOP
HANDHOLE
1"C, 3#10 TO MCC-1
FOR CONTINUATION SEE DWG NO. SK-E-108

AREA FLOODLIGHT POLE
DETAIL "B"
SK-E-108

## GENERAL NOTES

1. FOR A COMPLETE LIST OF ELECTRICAL REFERENCE DRAWINGS SEE DWG. NO. SK-E-100.

150 W HPS STANCHION MOUNT FIXTURE C-H CAT # VMS J15 OGP
1 1/4" CONDUIT STANCHION
1" COUPLING WELDED IN STANCHION
FLOOR FLANGE
PUMP PAD
TO CONTACTOR
1 1/4" CONDUIT STANCHION
DRILL 1 1/4" CONDUIT AND WELD IN 1" COUPLING
1" "L" FITTING
1" NIPPLE

STANCHION MOUNT
DETAIL "C"
SK-E-108

PIPERACK STEEL
P-1000 UNISTRUT WELDED TO UNDERNEATH SIDE OF RACK STEEL.
C-H UNY 205
UNISTRUT SPRING BOLT & NUT CAT # P-2379 & HFLW 031EG
200 W 120 V HPS CEILING MOUNT FIXTURE W/GLOBE & GUARD C-H CAT # VMVS 3C15 OGP

TYPICAL RACK FIXTURE MOUNTING
DETAIL "D"
SK-E-108

## REFERENCE DRAWINGS

SK-E-100 CLASSIFICATION & KEY PLAN
SK-E-108 AREA LIGHTING PLAN

PIPERACK STEEL
P-1000 UNISTRUT WELDED TO PIPERACK STEEL LENGTH AS REQUIRED.
UNISTRUT CONDUIT CLAMP TYPE P-2909 THRU P-2197 SIZE AS REQUIRED.

CONDUIT SUPPORT
DETAIL "E"
SK-E-108

2" 2" 2"
1 1/2"
2"
1/4" ϕ HOLES FOR 1 1/2" U-BOLT
2"
MATERIAL:
1/4"x2"x10" PL GALV.

STANDOFF
DETAIL "F"
THIS DRAWING

| DR | DATE | JOB NO. 12345 | ELECTRICAL DRAFTING MANUAL | | |
|---|---|---|---|---|---|
| CK | DATE | SECT. NO. | LIGHTING DETAILS | | |
| | | | SCALE NONE | DWG. NO. SK-E-114 | REV. A |

FIGURE 10.19.

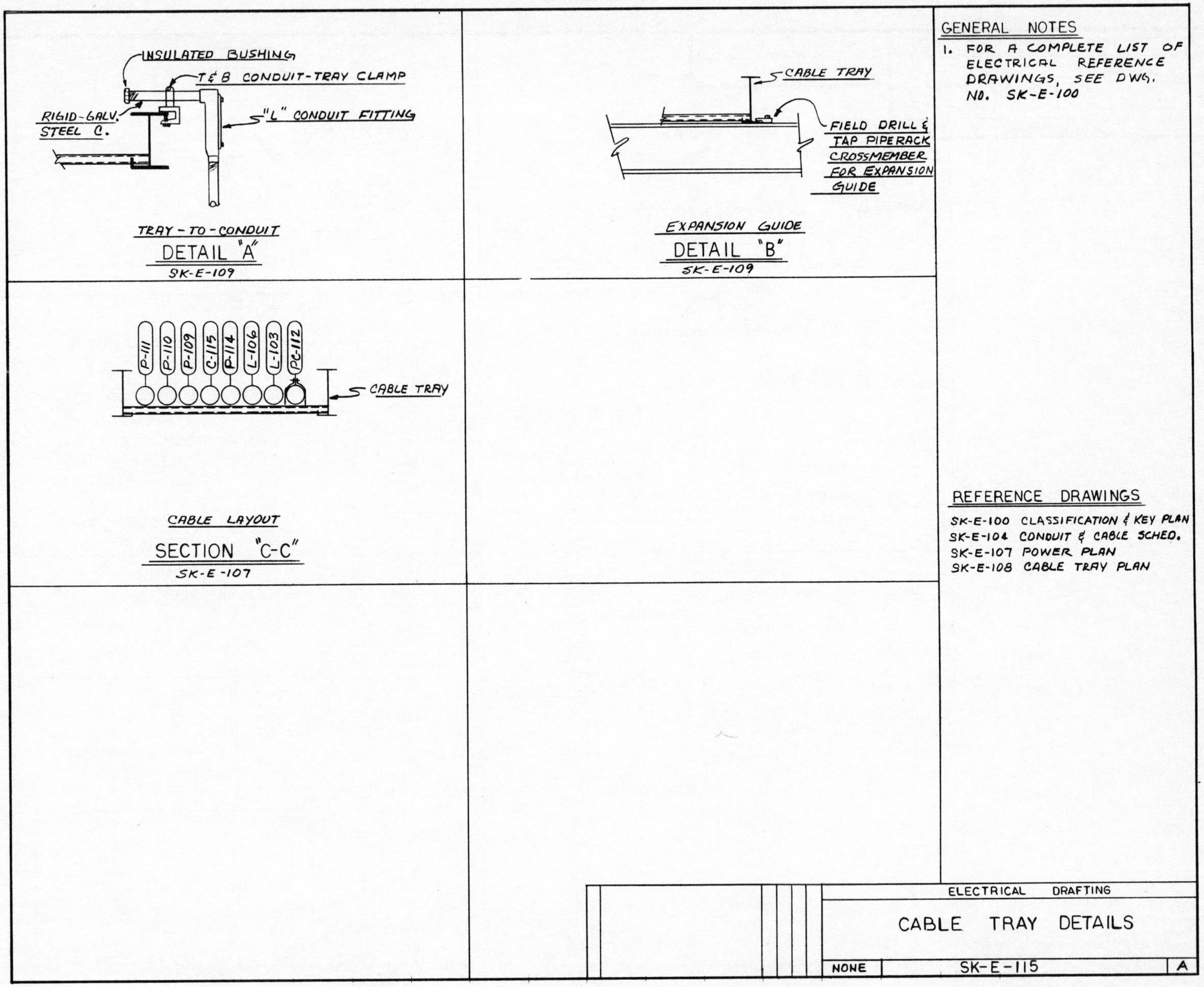

INSULATED BUSHING
T&B CONDUIT-TRAY CLAMP
RIGID-GALV. STEEL C.
"L" CONDUIT FITTING
TRAY-TO-CONDUIT
DETAIL "A"
SK-E-109
CABLE TRAY
FIELD DRILL & TAP PIPERACK CROSSMEMBER FOR EXPANSION GUIDE
EXPANSION GUIDE
DETAIL "B"
SK-E-109
P-111
P-110
P-109
C-115
P-114
L-106
L-103
PC-112
CABLE TRAY
CABLE LAYOUT
SECTION "C-C"
SK-E-107
GENERAL NOTES
1. FOR A COMPLETE LIST OF ELECTRICAL REFERENCE DRAWINGS, SEE DWG. NO. SK-E-100
REFERENCE DRAWINGS
SK-E-100 CLASSIFICATION & KEY PLAN
SK-E-104 CONDUIT & CABLE SCHED.
SK-E-107 POWER PLAN
SK-E-108 CABLE TRAY PLAN
ELECTRICAL DRAFTING
CABLE TRAY DETAILS
NONE
SK-E-115
A


**FIGURE 10.20.**

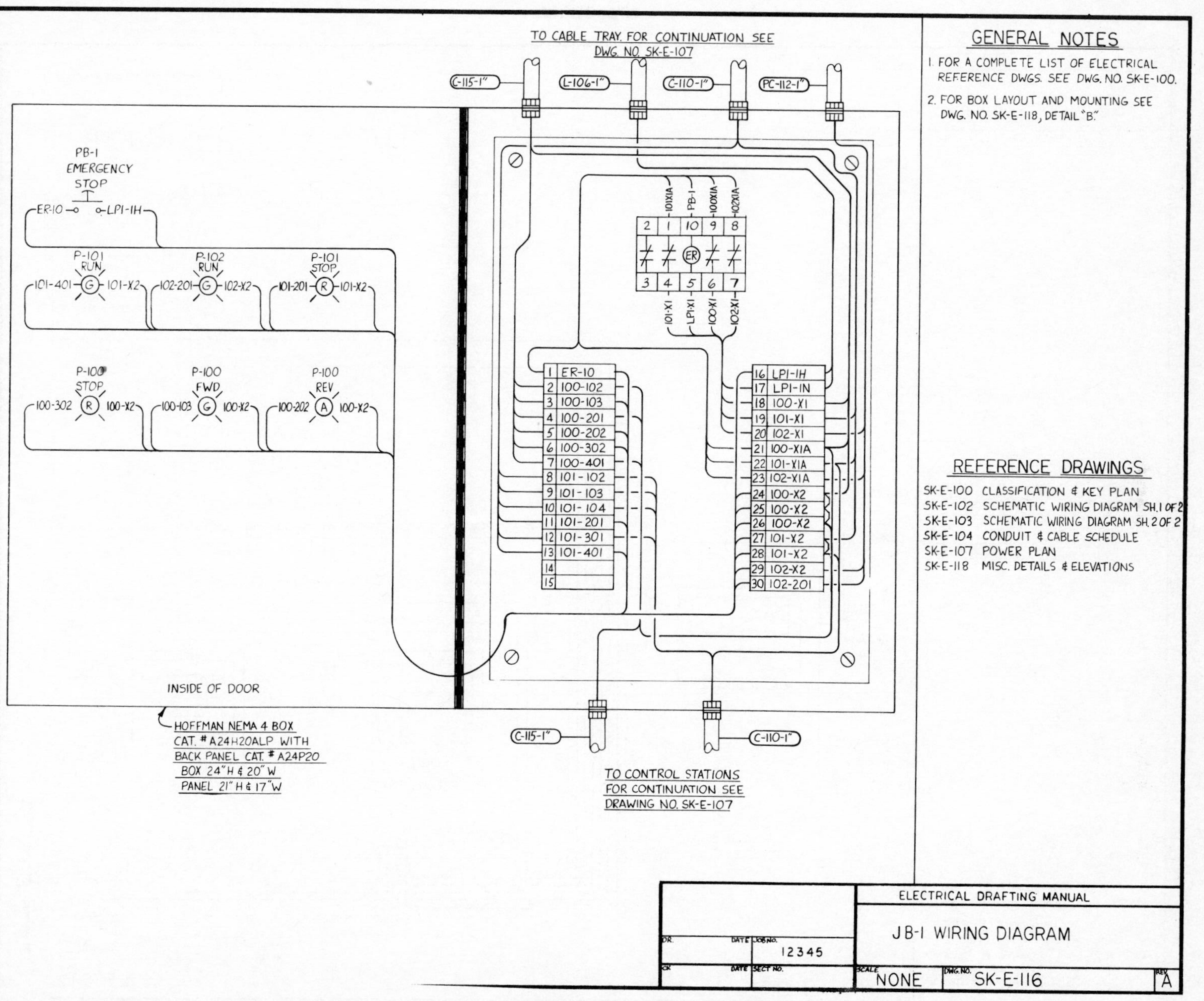

TO CABLE TRAY. FOR CONTINUATION SEE DWG. NO. SK-E-107
C-115-1"
L-106-1"
C-110-1"
PC-112-1"
GENERAL NOTES
1. FOR A COMPLETE LIST OF ELECTRICAL REFERENCE DWGS. SEE DWG. NO. SK-E-100.
2. FOR BOX LAYOUT AND MOUNTING SEE DWG. NO. SK-E-118, DETAIL "B".
PB-1 EMERGENCY STOP
ER-10
LP1-1H
P-101 RUN
101-401
101-X2
P-102 RUN
102-201
102-X2
P-101 STOP
101-201
101-X2
P-100 STOP
100-302
100-X2
P-100 FWD
100-103
100-X2
P-100 REV
100-202
100-X2
101X1A
PB-1
100X1A
102X1A
2 1 10 9 8
ER
3 4 5 6 7
101-X1
LP1-X1
100-X1
102-X1
1 ER-10
2 100-102
3 100-103
4 100-201
5 100-202
6 100-302
7 100-401
8 101-102
9 101-103
10 101-104
11 101-201
12 101-301
13 101-401
14
15
16 LP1-1H
17 LP1-1N
18 100-X1
19 101-X1
20 102-X1
21 100-X1A
22 101-X1A
23 102-X1A
24 100-X2
25 100-X2
26 100-X2
27 101-X2
28 101-X2
29 102-X2
30 102-201
REFERENCE DRAWINGS
SK-E-100 CLASSIFICATION & KEY PLAN
SK-E-102 SCHEMATIC WIRING DIAGRAM SH. 1 OF 2
SK-E-103 SCHEMATIC WIRING DIAGRAM SH. 2 OF 2
SK-E-104 CONDUIT & CABLE SCHEDULE
SK-E-107 POWER PLAN
SK-E-118 MISC. DETAILS & ELEVATIONS
INSIDE OF DOOR
HOFFMAN NEMA 4 BOX CAT. # A24H20ALP WITH BACK PANEL CAT. # A24P20
BOX 24" H & 20" W
PANEL 21" H & 17" W
C-115-1"
C-110-1"
TO CONTROL STATIONS FOR CONTINUATION SEE DRAWING NO. SK-E-107
ELECTRICAL DRAFTING MANUAL
JB-1 WIRING DIAGRAM
DR.
DATE
JOB NO.
12345
CK
DATE
SECT NO.
SCALE NONE
DWG. NO. SK-E-116
REV A


FIGURE 10.21.

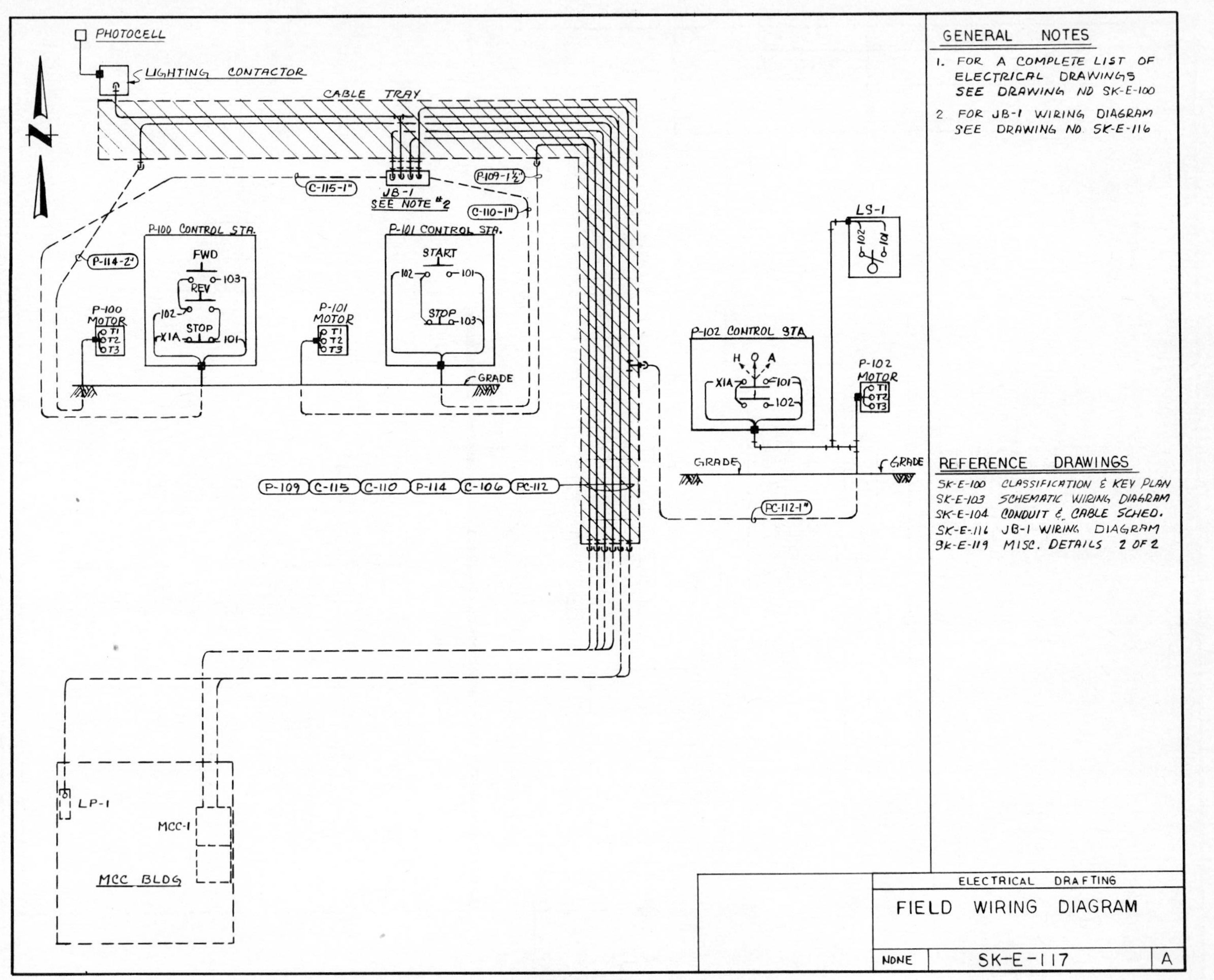

PHOTOCELL
LIGHTING CONTACTOR
CABLE TRAY
JB-1
SEE NOTE #2
C-115-1"
P-109-1½"
C-110-1"
P-114-2"
P-100 CONTROL STA.
FWD
REV
STOP
P-101 CONTROL STA.
START
STOP
P-100 MOTOR
P-101 MOTOR
GRADE
P-102 CONTROL STA.
H O A
LS-1
P-102 MOTOR
P-109
C-115
C-110
P-114
C-106
PC-112
PC-112-1"
LP-1
MCC-1
MCC BLDG
GENERAL NOTES
1. FOR A COMPLETE LIST OF ELECTRICAL DRAWINGS SEE DRAWING NO SK-E-100
2 FOR JB-1 WIRING DIAGRAM SEE DRAWING NO. SK-E-116
REFERENCE DRAWINGS
SK-E-100 CLASSIFICATION & KEY PLAN
SK-E-103 SCHEMATIC WIRING DIAGRAM
SK-E-104 CONDUIT & CABLE SCHED.
SK-E-116 JB-1 WIRING DIAGRAM
SK-E-119 MISC. DETAILS 2 OF 2
ELECTRICAL DRAFTING
FIELD WIRING DIAGRAM
NONE
SK-E-117
A


FIGURE 10.22.

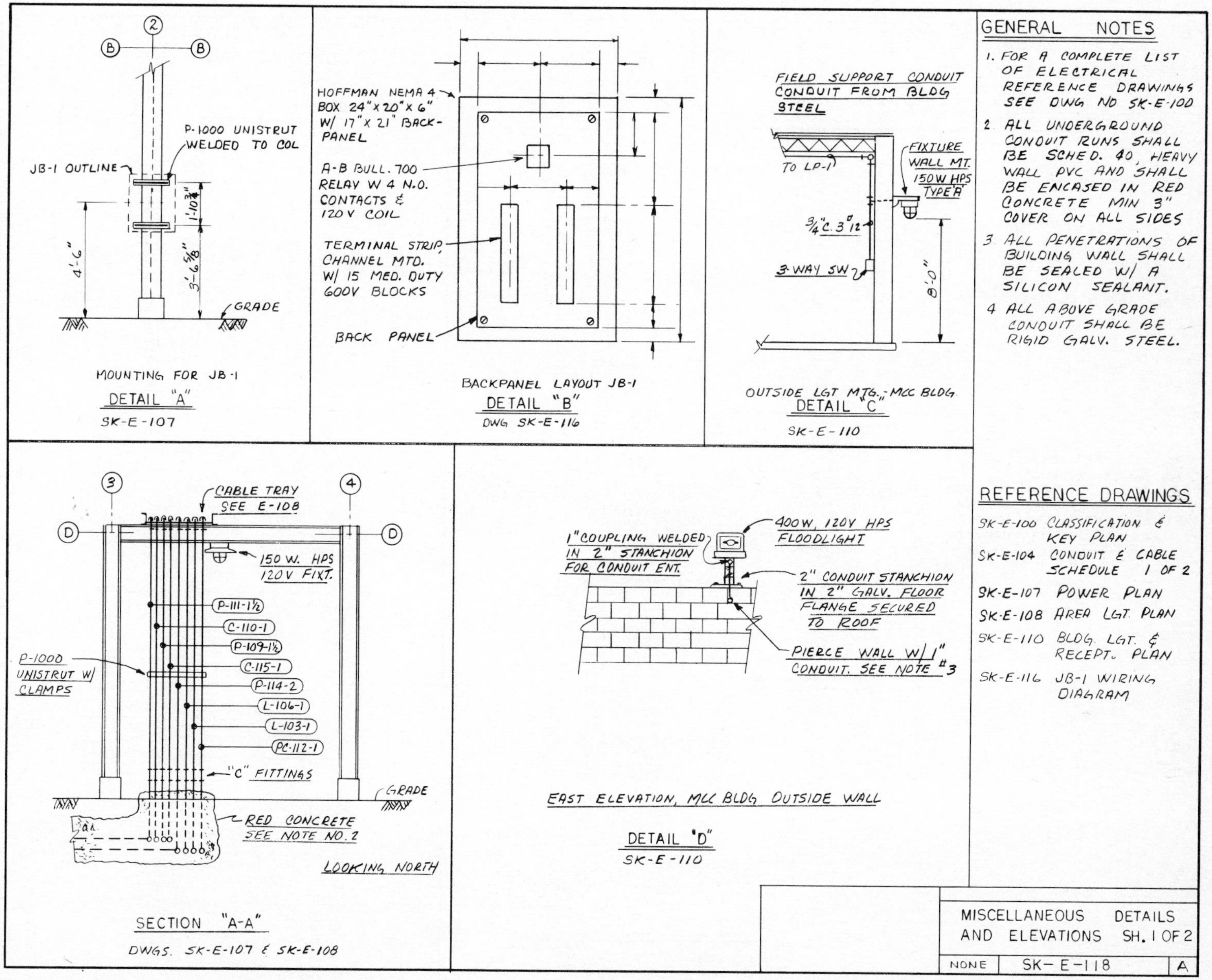

MOUNTING FOR JB-1

DETAIL "A"

SK-E-107

JB-1 OUTLINE

P-1000 UNISTRUT WELDED TO COL

1'-10¾"

4'-6"

3'-6⅝"

GRADE

HOFFMAN NEMA 4 BOX 24" x 20" x 6" W/ 17" x 21" BACK-PANEL

A-B BULL. 700 RELAY W 4 N.O. CONTACTS & 120 V COIL

TERMINAL STRIP, CHANNEL MTD. W/ 15 MED. DUTY 600V BLOCKS

BACK PANEL

BACKPANEL LAYOUT JB-1

DETAIL "B"

DWG SK-E-116

FIELD SUPPORT CONDUIT CONDUIT FROM BLDG STEEL

TO LP-1

FIXTURE WALL MT. 150W HPS TYPE 'A'

¾"C. 3 #12

3-WAY SW

8'-0"

OUTSIDE LGT MTG. - MCC BLDG

DETAIL "C"

SK-E-110

## GENERAL NOTES

1. FOR A COMPLETE LIST OF ELECTRICAL REFERENCE DRAWINGS SEE DWG NO SK-E-100
2. ALL UNDERGROUND CONDUIT RUNS SHALL BE SCHED. 40, HEAVY WALL PVC AND SHALL BE ENCASED IN RED CONCRETE MIN 3" COVER ON ALL SIDES
3. ALL PENETRATIONS OF BUILDING WALL SHALL BE SEALED W/ A SILICON SEALANT.
4. ALL ABOVE GRADE CONDUIT SHALL BE RIGID GALV. STEEL.

CABLE TRAY SEE E-108

150 W. HPS 120V FIXT.

P-111-1½

C-110-1

P-109-1½

C-115-1

P-114-2

L-106-1

L-103-1

PC-112-1

P-1000 UNISTRUT W/ CLAMPS

"C" FITTINGS

GRADE

RED CONCRETE SEE NOTE NO. 2

LOOKING NORTH

SECTION "A-A"

DWGS. SK-E-107 & SK-E-108

1" COUPLING WELDED IN 2" STANCHION FOR CONDUIT ENT.

400W, 120V HPS FLOODLIGHT

2" CONDUIT STANCHION IN 2" GALV. FLOOR FLANGE SECURED TO ROOF

PIERCE WALL W/ 1" CONDUIT. SEE NOTE #3

EAST ELEVATION, MCC BLDG OUTSIDE WALL

DETAIL "D"

SK-E-110

## REFERENCE DRAWINGS

SK-E-100 CLASSIFICATION & KEY PLAN

SK-E-104 CONDUIT & CABLE SCHEDULE 1 OF 2

SK-E-107 POWER PLAN

SK-E-108 AREA LGT. PLAN

SK-E-110 BLDG. LGT. & RECEPT. PLAN

SK-E-116 JB-1 WIRING DIAGRAM

| MISCELLANEOUS DETAILS AND ELEVATIONS SH. 1 OF 2 | | |
|---|---|---|
| NONE | SK-E-118 | A |

FIGURE 10.23.

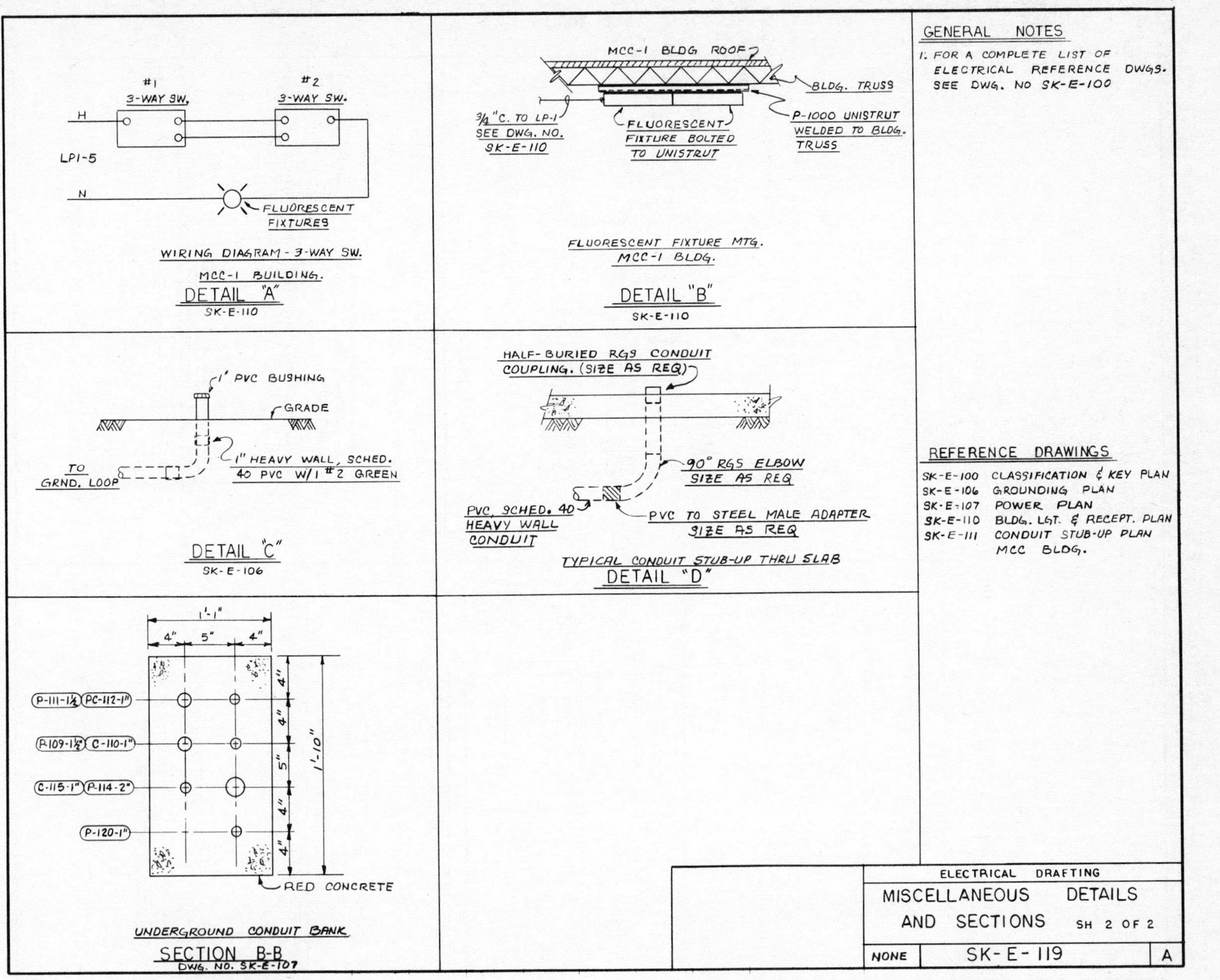
GENERAL NOTES
1. FOR A COMPLETE LIST OF ELECTRICAL REFERENCE DWGS. SEE DWG. NO SK-E-100
#1 3-WAY SW.
#2 3-WAY SW.
H
LP1-5
N
FLUORESCENT FIXTURES
WIRING DIAGRAM - 3-WAY SW.
MCC-1 BUILDING.
DETAIL "A"
SK-E-110
MCC-1 BLDG ROOF
BLDG. TRUSS
3/4" C. TO LP-1 SEE DWG. NO. SK-E-110
FLUORESCENT FIXTURE BOLTED TO UNISTRUT
P-1000 UNISTRUT WELDED TO BLDG. TRUSS
FLUORESCENT FIXTURE MTG.
MCC-1 BLDG.
DETAIL "B"
SK-E-110
1" PVC BUSHING
GRADE
TO GRND. LOOP
1" HEAVY WALL, SCHED. 40 PVC W/1 #2 GREEN
DETAIL "C"
SK-E-106
HALF-BURIED RGS CONDUIT COUPLING. (SIZE AS REQ)
90° RGS ELBOW SIZE AS REQ
PVC, SCHED. 40 HEAVY WALL CONDUIT
PVC TO STEEL MALE ADAPTER SIZE AS REQ
TYPICAL CONDUIT STUB-UP THRU SLAB
DETAIL "D"
REFERENCE DRAWINGS
SK-E-100 CLASSIFICATION & KEY PLAN
SK-E-106 GROUNDING PLAN
SK-E-107 POWER PLAN
SK-E-110 BLDG. LGT. & RECEPT. PLAN
SK-E-111 CONDUIT STUB-UP PLAN MCC BLDG.
1'-1"
4"
5"
4"
1'-10"
P-111-1½
PC-112-1"
P-109-1½
C-110-1"
C-115-1"
P-114-2"
P-120-1"
RED CONCRETE
UNDERGROUND CONDUIT BANK
SECTION B-B
DWG. NO. SK-E-107
ELECTRICAL DRAFTING
MISCELLANEOUS DETAILS AND SECTIONS
SH 2 OF 2
NONE
SK-E-119
A

**FIGURE 10.24.**

# AREA CLASSIFICATION

## 11.1 DEFINITION

Area classification is the determination of the fire or explosion hazards that may exist due to flammable gases or vapors, flammable liquids, combustible dust, or ignitible fibers or flyings. Locations are classified depending on the properties of the flammable vapors, liquids, or gases and combustible dusts or fibers that may be present and the possibility that a flammable or combustible concentration or quantity may be or is present. In determining classification, it is necessary to consider each room, section, or area on an individual basis.

## 11.2 FACTORS IN INITIAL DESIGN

Determinations of the product involved and the category under which it falls are important decisions in initial design. From the determination of hazardous (classified) locations, it is frequently possible to locate much of the equipment in less hazardous or even in nonhazardous areas and thus reduce the amount of special equipment required and the overall project cost. In some cases, hazards may be reduced by adequate positive-pressure ventilation from a source of clean air, which will purge or pressurize the enclosure or building.

Another method of reducing hazards from arcing devices is the use of intrinsically safe devices and equipment where the portion that presents a hazard, the arcing device, is hermetically sealed to prevent the passage of gases or vapors.

## 11.3 CODE DEFINITIONS

At this point, there are two definitions to consider:

*Approved:* Acceptable to the authority having jurisdiction.

*Explosion-proof apparatus:* Apparatus enclosed in a case that is capable of withstanding an explosion of a specified gas or vapor that may occur within it and of preventing the ignition of a specified gas or vapor surrounding the enclosure by sparks, flashes, or explosion of the gas or vapor within, and which operates at such an external temperature that a surrounding flammable atmosphere will not be ignited.

As we continue our discussion of classified areas, we will be confronted with these two words, especially in the equipment and device discussions.

## 11.4 AREA LIMITS

Once the type of vapor or gas is determined and the classification and group established, it is necessary to establish area limits. In all practical cases, the outer limits of the area will be 50 ft in all directions from the source of the gas or vapor. This means that every piece of equipment, every device, and all enclosures that fall within these limits must be approved for this gas or vapor.

This information must be recorded, so we devise a plan designating the classification of the various areas symbolically and defining the limits of each area. This drawing is in the initial group of drawings laid out in any package. It is necessary to determine the type of equipment, devices, and the like necessary to meet the requirements of the area classification and to determine layout of conduit and equipment used in this hazardous area. See Figure 11.1.

## 11.5 CONDUIT SYSTEMS IN CLASSIFIED AREAS

The conduit systems used in classified areas must be threaded with an NPT standard cutting die that provides 3/4-in. taper per foot. Such conduit must be made up wrench tight to minimize sparking when fault current flows through the conduit system. Where it is impractical to make a threaded joint tight, a bonding jumper should be utilized.

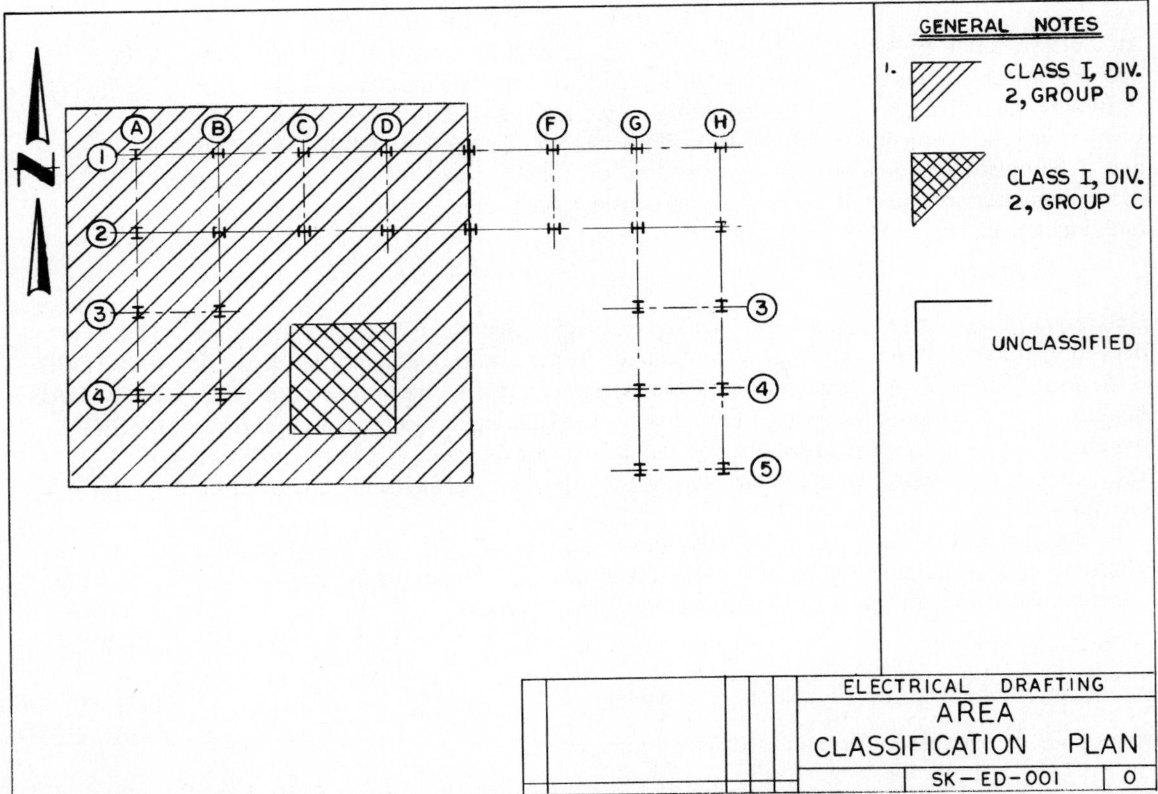


**FIGURE 11.1.** Area Classification

## 11.6 CLASS AND DIVISION DEFINITIONS

With the preceding in mind, we make the following definitions:

### 11.6.1 Class I Locations

Class I locations are those in which flammable gases or vapors are or may be present in the air in quantities sufficient to produce explosive or ignitable mixtures. Class I locations include those specified in the following.

### 11.6.2 Class I, Division 1

A Class I, Division 1 location is a location (1) in which ignitable concentrations of flammable gases or vapors exist under normal operating conditions, or (2) in which ignitable concentrations of such gases or vapors may exist frequently because of repair or maintenance operations or because of leakage, or (3) in which breakdown or faulty operation of equipment or processes might release ignitable concentrations of flammable gases or vapors and might also cause simultaneous failure of electric equipment.

This classification usually includes locations where volatile flammable liquids or liquified flammable gases are transferred from one container to another; interiors of spray booths and areas in the vicinity of spraying and painting operations where volatile flammable solvents are used; locations containing open tanks or vats of volatile flammable liquids; drying rooms or compartments for the evaporation of flammable solvents; locations containing fat and oil extraction equipment using volatile flammable solvents; portions of cleaning and dyeing plants where flammable liquids are used; gas generator rooms and other portions of gas manufacturing plants where flammable gas may escape; inadequately ventilated pump rooms for flammable gas or for volatile flammable liquids; the interiors of refrigerators and freezers in which volatile flammable materials are stored in open, lightly stoppered, or easily ruptured containers; and all other locations where ignitable concentrations of flammable vapors or gases are likely to occur in the course of normal operations.

### 11.6.3 Class I, Division 2

A Class I, Division 2 location is a location (1) in which volatile flammable liquids or flammable gases are handled, processed, or used, but in which the liquids, vapors, or gases will normally be confined within closed containers or closed systems from which they can escape only in case of accidental rupture or breakdown of such containers or systems, or in case of abnormal operation of equipment, or (2) in which ignitable concentrations of gases or vapors are normally prevented by positive mechanical ventilation, and which might become hazardous through failure of abnormal operation of the ventilating equipment; or (3) that is adjacent to a Class I, Division 1 location, and to which ignitable concentrations of gases or vapors might occasionally be communicated unless such communication is prevented by adequate positive-pressure ventilation from a source of clean air, and effective safeguards against ventilation failure are provided.

This classification usually includes locations where volatile flammable liquids or flammable gases or vapors are used, both which, in the judgment of the authority having jurisdiction, would become hazardous only in case of an accident or of some unusual operating condition. The quantity of flammable material that might escape in case of accident, the adequacy of ventilating equipment, the total area involved, and the record of the industry or business with respect to explosions or fires are all factors that merit consideration in determining the classification and extent of each location.

Piping without valves, checks, meters, and similar devices would not ordinarily introduce a hazardous condition even though used for flammable liquids or gases. Locations used for the storage of flammable liquids or of liquified or compressed gases in sealed containers would not normally be considered hazardous unless subject to other hazardous conditions also.

Electrical conduits and their associated enclosures separated from process fluids by a single seal or barrier are classed as a Division 2 location if the outside of the conduit and enclosures is a nonhazardous location.

### 11.6.4 Class II Locations

Class II locations are those that are hazardous because of the presence of combustible dust. Class II locations include those specified in the following.

### 11.6.5 Class II, Division 1

A Class II, Division 1 location is a location (1) in which combustible dust is in the air under normal operating conditions in quantities sufficient to produce explosive or ignitable mixtures; or (2) where mechanical failure or abnormal operation of machinery or equipment might cause such explosive or ignitable mixtures to be produced, and might also provide a source of ignition through simultaneous failure of electric equipment, operation of protection devices, or from other causes; or (3) in which combustible dusts of an electrically conductive nature may be present.

Combustible dusts that are electrically nonconductive include dusts produced in the handling and processing of grain and grain products, pulverized sugar and cocoa, dried egg and milk powders, pulverized spices, starch, and pastes, potato and wood flour, oil and meal from beans and seed, dried hay, and other organic materials that may produce combustible dusts when processed or handled.

Dusts containing magnesium or aluminum are particularly hazardous, and the use of extreme precaution is necessary to avoid ignition and explosion.

### 11.6.6 Class II, Division 2

A Class II, Division 2 location is a location in which (1) combustible dust will not normally be in suspension in the air in quantities sufficient to produce explosive or ignitable mixtures, and dust accumulations are normally insufficient to interfere with the normal operation of electrical equipment or other apparatus, or (2) dust may be in suspension in the air as a result of infrequent malfunctioning of handling of processing equipment, and dust accumulations resulting therefrom may be ignitable by abnormal operation or failure of electrical equipment or other apparatus.

### 11.6.7 Class III Locations

Class III locations are those that are hazardous because of the presence of easily ignitable fibers or flyings, but in which such fibers or flyings are not likely to be in suspension in the air in quantities sufficient to produce ignitable mixtures. Class III locations include those specified in the following:

### 11.6.8 Class III, Division 1

A Class III, Division 1 location is a location in which easily ignitable fibers or materials producing combustible flyings are handled, manufactured, or used. Such locations usually include some parts of rayon, cotton, and other textile mills; combustible fiber manufacturing and processing plants; cotton gins and cottonseed mills; flax-processing plants; clothing manufacturing plants; sawmills and other woodworking locations; and establishments and industries involving similar hazardous processes or conditions.

Easily ignitable fibers and flyings include rayon, cotton (including cotton linters and cotton waste), sisal or henequen, istle, jute, hemp, tow, cocoa fiber, oakum, baled waste kapok, Spanish moss, excelsior, sawdust, wood chips, and other materials of similar nature.

### 11.6.9 Class III, Division 2

A Class III, Division 2 location is a location in which easily ignitable fibers are stored or handled (not in the process of manufacture).

### 11.6.10 NEMA Enclosure Class and Group Designations

*Class I:* Enclosures for flammable volatile liquids and highly flammable gases that form explosive mixtures. Must be suitably rigid to withstand four times the pressure of an internal explosion that may be caused should gases enter enclosure. All joints, including shafts extending through the enclosure, must have sufficiently long flame path at close tolerances to prevent any flame from an internal explosion from being transmitted to the outside of the enclosure. Use of gasket not permitted.

*Class I, Group A:* Atmospheres containing acetylene.

*Class I, Group B:* Atmospheres containing hydrogen, gases, or vapors of equivalent hazard, such as manufactured gases containing more than 30 percent hydrogen by volume.

*Class I, Group C:* Atmospheres containing ethyl ether vapor.

*Class I, Group D:* Atmospheres containing gasoline, petroleum, naphtha, alcohols, acetone, lacquer solvent vapor, or natural gas.

*Class II:* Enclosures for preventing explosive mixtures of dust and air from coming into contact with hot or arcing parts or equipment that must not become overheated when operating under a dust blanket. Enclosure must have nonrubber gasket to keep dust out and does not have to be able to withstand an internal explosion.

*Class II, Group E:* Atmospheres containing metal dust.

*Class II, Group F:* Atmospheres containing carbon black, coal, or coke gas.

*Class II, Group G:* Atmospheres containing grain dust.

### 11.6.11 NEMA Hazardous Location Designations

*Class I, Division 1:* Hazardous concentrations of flammable gases or vapors exist continuously, intermittently, or periodically under normal operating conditions.

*Class I, Division 2:* Hazardous concentrations of flammable gases or vapors are handled in closed containers or systems.

*Class II, Division 1:* Dust is suspended in air continuously, intermittently, or periodically under normal operating conditions in quantities sufficient to produce explosive or ignitable mixtures.

**SELF-CHECK QUIZ 11.1** **Cover the right side of the page and answer the questions.**

| | |
|---|---|
| 11.1 The determination of fire and explosion hazards that exist in an area is called ____________ ____________. | 11.1 Ans.: *area classification* Ref. 11.1 |
| 11.2 ____________ means acceptable to the authority having jurisdiction. | 11.2 Ans.: *Approved* Ref. 11.3 |
| 11.3 Class I areas deal with ____________ ____________ ____________. | 11.3 Ans.: *gases and vapors* Ref. 11.6.1 |
| 11.4 A location in which ignitable concentrations exist at all times is a division ____________location. | 11.4 Ans.: *1* Ref. 11.6.2 |
| 11.5 A location in which ignitable concentrations are contained but may be present accidentally is a division ____________area. | 11.5 Ans.: *2* Ref. 11.6.3 |
| 11.6 Class II areas deal with ____________. | 11.6 Ans.: *dust* Ref. 11.6.4 |
| 11.7 Class III areas deal with ____________ or ____________. | 11.7 Ans.: *fibers* or *flyings* Ref. 11.6.7 |
| 11.8 Explosion-proof equipment is designed to ____________an explosion. | 11.8 Ans.: *contain* Ref. 11.3 |

*Class II, Division 2:* Dust is not suspended, but accumulates on electrical equipment sufficient to prohibit sufficient cooling, causing a fire hazard.

*Class III, Division 1:* Ignitable fibers or materials producing combustible flyings are being handled, manufactured, or used.

*Class III, Division 2:* Ignitable fibers are being stored or handled (other than in process of manufacture).

# 12 COMMERCIAL AND RESIDENTIAL DRAWINGS

**12.1 GENERAL** The same basic principles of drafting that apply to projects in an industrial environment also apply to the field of commercial and residential drafting. There are, however, some distinct practices and principles in commercial electrical drafting that have a tendency to set it aside as a separate field of endeavor. To better understand these differences, a comparison of industrial versus commercial–residential wiring practices is set forth in Table 12.1.

Table 12.2 provides the symbols used in commercial–residential drafting and is followed by sample drawings in Figures 12.1 and 12.2.

**Table 12.1 ELECTRICAL DRAFTING: INDUSTRIAL COMPARED TO COMMERCIAL–RESIDENTIAL**

| *Item* | *Industrial* | *Commercial–Residential* |
|---|---|---|
| Voltages | Industrial installations utilize a wide variety of voltages, especially heavy industry. High voltages in the 5 to 15-kV range are used as supply voltages and stepped down for branch circuits within the plant. Three-phase circuitry is predominant, although single phase is also used. Direct current in 6, 12, and 24-V ranges is used for instrumentation. | The majority of commercial and residential installations will be 120/240 single-phase service. Sometimes a 480-V service is used and transformed to 120/240. Only in large commercial projects are higher service voltages used. |
| Symbols | The majority of symbols used in industrial drafting are also applicable to commercial–residential with some exceptions applying to special equipment, instrumentation, and/or electronics. | Uses many symbols used in industrial drafting but also has a number that are only used in commercial–residential, due to the many devices that are depicted only in this area. See symbol list at end of this chapter. |
| Items or devices | Most items and devices used can be represented by a symbol, either standard or special for an installation. | Many special devices used only in commercial–residential drafting require symbols such as residential signaling systems and institutional occupancies. See symbol list at end of this chapter. |
| Wiring | In the majority, cable and/or conduit routing as well as wiring is critical in an industrial complex. Due to the large volume of wires and the distance involved, wire numbers and markers are generally required. Wiring and raceway routing are generally tied down in industrial drafting. More physical protection against damage to the wiring must be made. | Commercial–residential wiring in a majority of instances is "open wiring," not in raceways, and the routing from device to source is not critical. It usually is run in false ceilings or walls and many times not in conduit. Home runs are indicated by an arrow that does not specify a specific path to follow. See Figures 12.1 and 12.2. |
| Drawings and presentation | Industrial installation will generally require a greater number of drawings due to the complexity of the system. Specific routing, details, system tie-ins, instrumentation, and the many control systems and interlocks require explanations. In addition, the product, if hazardous, will increase the detail required. | Since the intricate controls are usually not present in commercial–residential, the number of wiring diagrams, conduit plans, plot plans, schematic diagrams, and the like, is greatly reduced. Riser diagrams can replace several plan drawings. Few details are drawn. |
| Theory | Electrical theory in service, circuitry, and control applies to all areas of drafting. Load requirements, formulas, calculations, and so on, apply equally. | See industrial description. |
| Codes | The National Electric Code, local codes and restrictions, and state and federal requirements apply to all areas of drafting equally. They will usually require closer compliance in an industrial complex, depending on product, than in commercial or residential, but safety is still paramount. | See industrial description. |

## Table 12.2a ARCHITECTURAL SYMBOLS

| Architectural Plans | |
|---|---|
| **Bus Ducts and Wireways** | **Panelboards** |
| T T T Trolley duct | Flush mounted panelboard and cabinet |
| B B B Busway (service, feeder, or plug-in) | Surface-mounted panelboard and cabinet |
| C C C Cable through ladder or channel | |
| W W W Wireway | |
| | |
| | |

| Architectural Plans | |
|---|---|
| **Electrical Distribution or Lighting System, Aerial** | |
| Pole | Head guy |
| Street light and bracket | Sidewalk guy |
| Transformer | Service weather |
| Primary circuit | |
| Secondary circuit | |
| Down guy | |

(a)

## Table 12.2b ARCHITECTURAL SYMBOLS

| Architectural Plans | |
|---|---|
| **Electrical Distribution or Lighting System, Underground** | |
| M Manhole | Street light standard fed from underground circuit |
| H Handhole | |
| TM Transformer-manhole or vault | |
| TP Transformer pad | |
| Underground direct burial cable | |
| Underground duct line | |

| Architectural Plans | |
|---|---|
| **Panel Circuits and Miscellaneous** | |
| Lighting panel | Home run to panel<br>Indicate number of circuits by number of arrows. Any circuit without such designation indicates a two-wire circuit. For a greater number of wires indicate as follows:<br>(3 wires)<br>(4 wires) etc. |
| Power panel | |
| Wiring, concealed in wall or ceiling | Wiring turned up |
| Wiring concealed in floor | Wiring turned down |
| Wiring exposed | G Generator |
| Feeder<br>Use heavy lines and designate by number corresponding to listing in feeder schedule | M Motor |

(b)

## Table 12.2c ARCHITECTURAL SYMBOLS

| Architectural Plans | |
|---|---|
| Panel Circuits and Miscellaneous | |
| I Instrument (specify) | |
| Controller | |
| Externally operated disconnect switch | |
| T Transformer (or draw to scale) | |
| | |
| | |

| Architectural Plans | |
|---|---|
| Signaling System Outlets Residential Occupancies | |
| Push button | Interconnecting telephone |
| Buzzer | Outside telephone |
| Bell | BT Bell-ringing transformer |
| Bell and buzzer combination | D Electric door opener |
| Annunciator | M Maid's signal plug |
| Telephone switchboard | R Radio outlet |

(c)

## Table 12.2d ARCHITECTURAL SYMBOLS

| Architectural Plans | |
|---|---|
| Signaling System Outlets Residential Occupancies | |
| CH Chime | |
| TV Television outlet | |
| T Thermostat | |
| | |
| | |
| | |

| Architectural Plans | |
|---|---|
| Institutional, Commercial, and Industrial Occupancies | |
| Nurses call system devices (any type) | Private telephone system devices |
| Paging system devices (any time) | Watchman system devices |
| Fire alarm system devices (any type) | Sound system |
| Staff register system (any time) | Other signal system devices |
| Electrical clock system devices (any type) | SC Signal central station |
| Public telephone system devices | Interconnection box |

(d)

## Table 12.2e ARCHITECTURAL SYMBOLS

| Architectural Plans | |
|---|---|
| Institutional, Commercial, and Industrial Occupancies | |
| — - — - — Auxiliary System Circuits<br>Any line without further designation indicates two-wire system. For a greater number of wires, designate with numerals in manner similar to 12-no 18 W-$\frac{3}{4}$″ C. Designate by numbers corresponding to the listing in schedule | |
| ☐ A, B, C, etc.<br>Subscript lettering refers to notes on drawings or detailed description in specifications. | |
| | |
| | |
| | |

| Architectural Plans | |
|---|---|
| Lighting Outlets | |
| Ceiling Wall | Ceiling Wall |
| Surface incandescent | (J) Junction box |
| (R) Recess incandescent | (L) PS Lamp holder with pull switch |
| (B) Blanked outlet | (V) Outlet for vapor discharge lamp |
| (D) Drop cord | (X) Exit light outlet |
| (E) Electrical outlet | (XR) Recessed exit light outlet |
| (F) Fan outlet | (L) Outlet controlled by low-voltage switching when relay is installed in outlet box |

(e)

## Table 12.2f ARCHITECTURAL SYMBOLS

| Architectural Plans | |
|---|---|
| Lighting Outlets | |
| O Surface or pendent individual fluorescent fixture | |
| OR Recessed individual fluorescent fixture | |
| O Surface or pendent continuous row fluorescent fixture | |
| OR Recessed continuous row fluorescent fixture | |
| | |
| | |

| Architectural Plans | |
|---|---|
| Receptacle Outlets | |
| Single receptacle outlet | Single special-purpose receptacle outlet |
| Duplex receptacle outlet | Duplex special-purpose receptacle outlet |
| Triplex receptacle outlet | S Switched receptacle outlet |
| Quadruplex receptacle outlet | R Range outlet |
| Duplex receptacle outlet, split wired | DW Special-purpose connection |
| Triplex receptacle outlet, split wired | $X^u$ Multioutlet assembly |

(f)

## Table 12.2g ARCHITECTURAL SYMBOLS

| Architectural Plans | |
|---|---|
| Receptacle Outlets | |
| (C) Clock hanger receptacle | Floor telephone outlet, public |
| Floor outlet receptacle | Floor telephone outlet, private |
| (F) Fan hanger receptacle | Underfloor duct and junction box for triple, double, or single duct system as indicated by number of parallel lines |
| Floor single receptacle outlet | Cellular floor headed duct |
| Floor duplex receptacle outlet | |
| Floor special purpose outlet | |

| Architectural Plans | |
|---|---|
| Special Outlets | |
| A, B, C, etc.<br>A, B, C, etc.<br>Any standard symbol as given above with the addition of lowercase subscript lettering may be used to designate some special variation of standard equipment of particular interest in specific set of architectural plans.<br>When used they must be listed in the schedule of symbols on each drawing and if necessary further described in the specifications. | |
| | |
| | |

(g)

## Table 12.2h ARCHITECTURAL SYMBOLS

| Architectural Plans | | | |
|---|---|---|---|
| Switch Outlets | | Switch Outlets | |
| $\$$ | Single-pole switch | $\$_P$ | Switch and pilot lamp |
| $\$_2$ | Double-pole switch | $\$_{CB}$ | Circuit breaker |
| $\$_3$ | Three-way switch | $\$_{WCB}$ | Weatherproof circuit breaker |
| $\$_4$ | Four-way switch | $\$_{MC}$ | Momentary contact switch |
| $\$_D$ | Automatic door switch | $\$_{RC}$ | Remote-control switch |
| $\$_K$ | Key-operated switch | $\$_{WP}$ | Weatherproof switch |

| Architectural Plans | | | |
|---|---|---|---|
| Switch Outlets | | Switch Outlets | |
| $\$_F$ | Fused switch | $-\ominus_S$ | Switch and single receptacle |
| $\$_{WF}$ | Weatherproof fused switch | $=\!\ominus_S$ | Switch and double receptacle |
| $\$_L$ | Switch for low-voltage switching system | $\$$ A, B, C, etc. | Any standard switch symbol as given above with the addition of lowercase subscript lettering may be used to designate some special variation of standard equipment of particular interest in specific set of architectural plans. When used they must be listed in the schedule of symbols on each drawing and if necessary further described in the specifications. |
| $\$_{LM}$ | Master switch for low-voltage switching system | | |
| $\$_T$ | Time switch | | |
| (\$) | Ceiling pull switch | | |

(h)

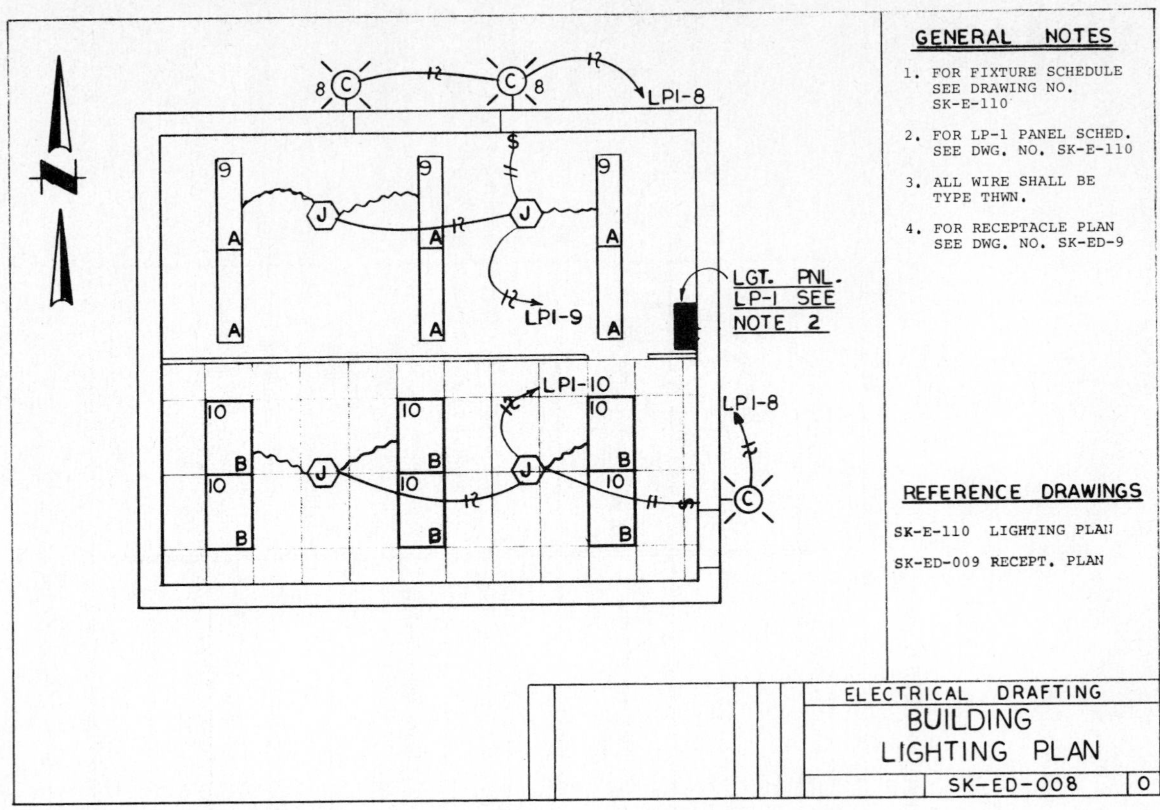


**FIGURE 12.1.** Commercial–Residential Drafting

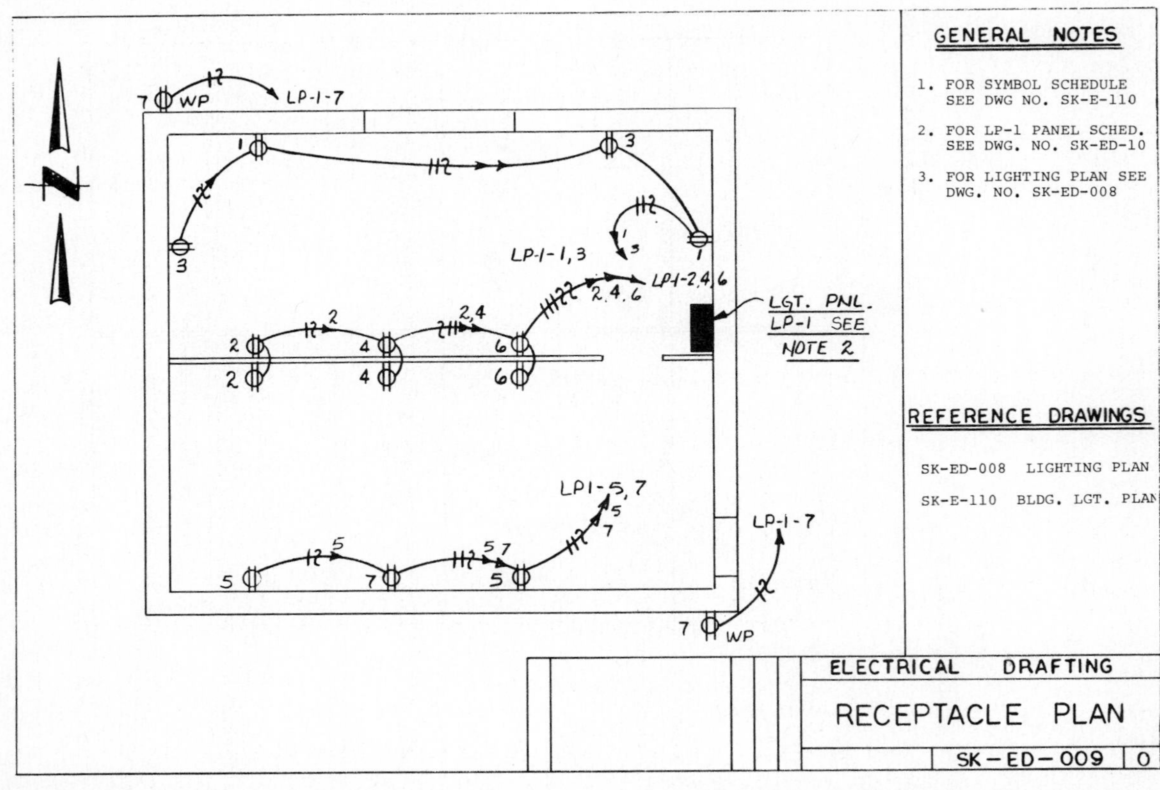


**FIGURE 12.2.** Commercial–Residential Drafting

# 13 COMPUTER GRAPHICS

## 13.1 GENERAL

Computers are having a profound effect on today's society. Our lives are not only influenced by these machines but in some cases are actually dependent upon them. Everyone knows that computers exist. We hear about them and read about them everyday. Our credit and charge account records are handled by computers; they oversee our banking and financial transactions; the airlines, hospitals, police, firemen, utility companies, and government offices all use computers in some form to conduct their daily business. To many people, a computer is a large, black, mysterious box isolated in a room, and some feel that the super intelligence displayed by computers will someday enable them to take over and direct our lives.

## 13.2 DEFINITION

In semi-technical language a computer could be defined as:

> *A device that receives, processes, and presents data*. It occurs in two types: analog and digital.

That is the basis of it. It is a machine that collects, memorizes, and processes large masses of coded information or data. These data are furnished by humans, in the form of programs, to the computer, which in turn stores it for recall. *The computer cannot think for itself*. It is entirely dependent on human help and most so-called computer errors are actually human errors, not computer errors. It can do no more than carry out the instructions of its operator.

## 13.3 EVOLUTION OF THE COMPUTER

Since the beginning of history, people have needed a method of counting. Primitive man used his fingers. Stone Age individuals used stones, sticks, and the like, and recorded numbers with notches on a stick. Around 3000 B.C. the abacus was developed by the

Chinese and, to this date, in the hands of a skilled operator, it is faster than most modern-day calculators.

Various other methods of calculating were developed through the years, such as logarithms in 1614, the slide rule in 1630, the first mechanical calculator in 1642, the first commercial calculator in 1820, and Univac 1, the world's first computer offered as a commercial product in 1951. The developments in computer technology since 1951 have brought us to the modern-day level of technology we live with and depend on.

## 13.4 BUSINESS AND THE COMPUTER

The computer has revolutionized the world of business. Technology has brought the price of small computers within the range of the bulk of small businesses. Where before many hours were spent on inventory, payables, receivables, and ledger items (expenses), it is now only a matter of minutes to recall this information, for it is stored in a computer memory bank and is subject to instant recall by merely pressing a button.

## 13.5 COMMUNICATING WITH THE COMPUTER

### 13.5.1 Symbols

The importance of symbols in our lives lies not in what they actually are but what we can do with them. The first set of symbols we become associated with is the alphabet. By themselves they mean nothing, but when rearranged and grouped we can form words with various meanings.

Later additional symbols, the numerals and punctuation marks, were programmed into our minds and we expanded our means and methods of communication. Similarly, we must communicate with the computer by using symbols that can be utilized by the computer equipment.

### 13.5.2 Cathode-ray Tube

For the past 25 to 30 years, the cathode-ray tube (CRT) has been associated with computers and display modules, first being used in oscilloscopes and readers, later in television, and finally in computers. The CRT terminal is a desk top unit that has the ability to allow one to enter as well as display data. The tube in your television set is a CRT that, through the electronics of the set, receives and displays a signal from a distant station.

The display area on the tube contains approximately 1 million points that can be individually controlled. The lighting of several points in a row displays a line. Other points added to this show curves or other parts of a figure. With this method, by lighting selected points, graphs, charts, lines, letters, numbers, and an infinite number of symbols and other figures can be displayed. At this point you may begin to realize the unlimited combinations available to the user.

## 13.6 COMPUTER-AIDED DRAFTING (CAD)

### 13.6.1 Display Units

Display units have been available to the computer user for some time. The display unit shows the operator lines of data or information called up from the memory bank. The ability to individually control the points of the cathode-ray tube and to change, switch, remove, and add lighted points is a result of present-day technology.

### 13.6.2 Light Pen

To give the operator command and control over the various points, a device called a light pen is used. This is a photosensitive device that detects the presence of light on the display screen. The light pen can be used to pinpoint information already on the screen or to designate a point at which information is to appear. It can also be used to print characters or symbols drawn up from the memory or menu of the computer.

### 13.6.3 Commands

Since most drawings used in industry and construction are composed of straight or curved lines and symbols, it is possible to reproduce these drawings on the display screen. Rough sketches of schematics, one-lines, plot plans, and the like, can be entered on the screen, changes are made and symbols added as desired, and the computer will then take over and display the finished product. After checking, the command to print is given and a device known as a digital plotter makes the drawing under the control of the computer. The finished drawing is on Mylar and in India ink. In addition, the computer, calling on its menu, will print a list of materials to install the equipment depicted by the drawing.

### 13.6.4 Storage and Memory

Many industrial concerns, such as switchgear manufacturers, deal in a repetitive operation where 90 percent of their customers order a standard or similar product and, in many cases, an identical product. The information developed in programming the computer, the menu put into the memory bank, completed drawings that the computer has done, can all be stored in memory and recalled on request. Picture the advantages of this feature to a company that has to furnish drawings of their product to a customer. With minor changes to the title block, a simple command will produce a finished drawing in approximately one-tenth of the time it would take a professional drafter to do it, and usually with much more accuracy. For the customer requirement that deviates from the standard, the standard can be called up on the screen and, using the light pen or cursor, all necessary changes can be made in a matter of minutes instead of hours, and upon command the computer will print the "special order."

### 13.6.5 Additional Advantages

In addition to providing consistent uniform drawings on a fine-quality paper or vellum done in India ink, many computer systems offer the option of setting up the drawings in layers. Some systems can accommodate as many as 63 layers of information. Buildings can be developed from the basement up, with each floor or layer available for instant recall. Pipe racks can be entered, and all levels complete with piping, conduit, structural steel, and devices and equipment at every level recorded and entered for instant recall or print. The possibilities are practically unlimited in the types and variety of drawings that can be stored in a CAD system.

## 13.7 TYPICAL SYSTEM AT WORK

### 13.7.1 The System Does the Work

Under control of the console operators, the computer performs such operations as drawing symbols, moving displayed items, drawing lines, correcting line positions, and copying portions of the displayed image. The operators, however, require absolutely no computer knowledge to control these and the many other operations performed durin; the preparation and modification of drawings. It is all done from the consoles with three easy-to-use devices: (1) a function keyboard, (2) an alphanumeric keyboard, and (3) a light pen. Figure 13.1 shows an operator at the console.

### 13.7.2 Function Keyboard

By pressing keys on the function keyboard, the operator selects the method used to modify and add information to the displayed image. In all there are 30 active keys on the average-function keyboard, including those for selecting such operations as draw, type, insert, and delete.

### 13.7.3 Alphanumeric Keyboard

With the alphanumeric keyboard, the operator types information on the CRT screen the same as typing on a sheet of paper in a typewriter. As each character is typed, a blinking underline, called the cursor, marks the position where the next character will appear.

### 13.7.4 Light Pen

Perhaps the most versatile control device available to the operator is the light pen. It can be used as a pointer for selecting various items and locations on the CRT screen and as a pen for drawing lines and positioning displayed data. A tracking cross appears on the CRT screen to mark the exact location selected with the light pen. The operator positions the tracking cross simply by moving the light pen across the face of the CRT.

### 13.7.5 Automatic Line Straightening and Positioning

Drawing with the light pen is almost like sketching with an ordinary pen, only it is easier, since the computer performs the job of keeping the lines straight. The operator aims the light pen at the point where he or she wants the line to start, and then moves it across the CRT screen to the line's termination point. As the operator does this, the computer

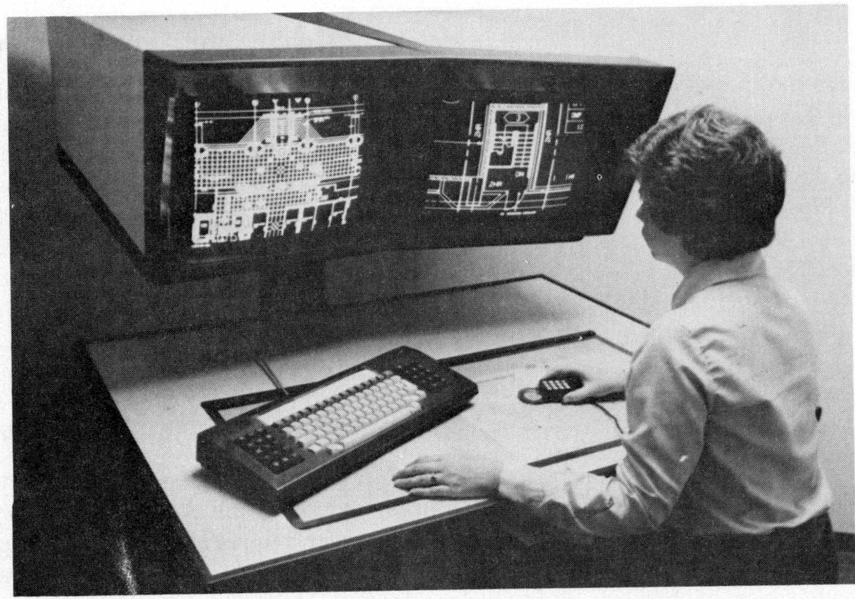

**FIGURE 13.1.** Operator at Console and Keyboard (Courtesy Engineering Graphics Technology, Houston, Texas)

draws a perfectly straight line between the two points, regardless of how erratically the operator moves the light pen.

By keying switches on the function keyboard, the operator can also select either of two modes of drawing lines. They are called MH and RB. Lines drawn in the MH mode are positioned so that they are parallel to the horizontal or vertical axis of the CRT screen. For example, if the operator attempts to draw a horizontal line but actually moves the light pen at an angle to the horizontal axis, the computer automatically repositions the line to its correct location. In the RB mode, the operator draws lines and then orients them in any direction by moving the light pen in an arc about the line's starting point. Figure 13.2 shows a typical command menu.

### 13.7.6 Types of Lines Drawn

Since it is often necessary to emphasize portions of drawings with various types of lines, the computer gives the operator a selection of dashed or solid lines and three line widths. These characteristics are displayed in an area at the bottom of the CRT called the *status block*. Using the light pen, the operator goes to the status block and points to the characteristics describing the type of line wanted. Each line drawn in the working area of the CRT will then exhibit these characteristics. The operator can change the line characteristics at will by returning to the status block and making another selection with the light pen.

### 13.7.7 Selection of Menu Symbols for Computer Drawings

The operator places symbols for transistors, logic elements, and other components on the drawing without having to perform the tedious job of tracing each on the CRT screen. The operator simply selects the name of the symbol from a list displayed in the status block, points to it with the light pen, and instantly the symbol appears on the CRT screen. The operator then uses the light pen to move the symbol to its correct position on the displayed drawing.

In all, there are over 180 symbols that the operator can select, including those

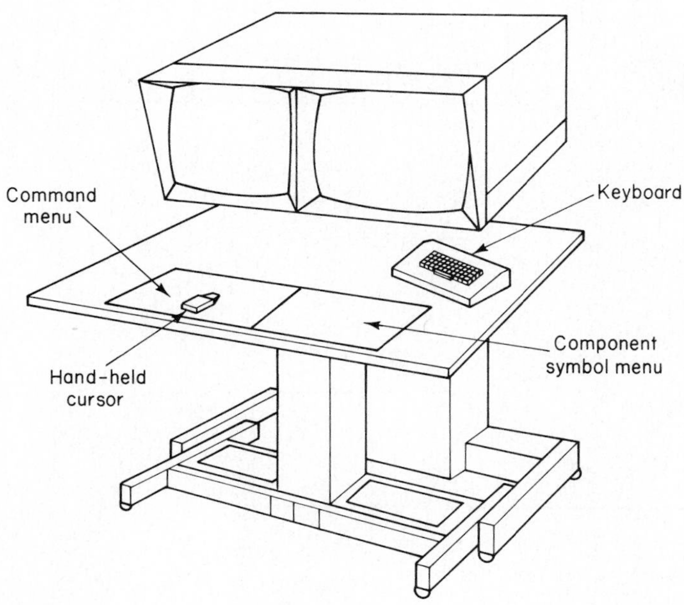


**FIGURE 13.2.** Operator Input Design Station (Courtesy Engineering Graphics Technology, Houston, Texas)

normally used in electrical, logical, mechanical, microwave, and hydraulic diagrams. With some symbols, the operator is given a display of characteristics from which to select the direction in which symbols are oriented and their size.

### 13.7.8 Additions, Deletions, and Changes Easily Made

Once the operator works up the graphic portions of the drawing, he or she uses the alphanumeric keyboard to type in component values, identification numbers, and other such information. If a typing error is made, the operator need only mark the position

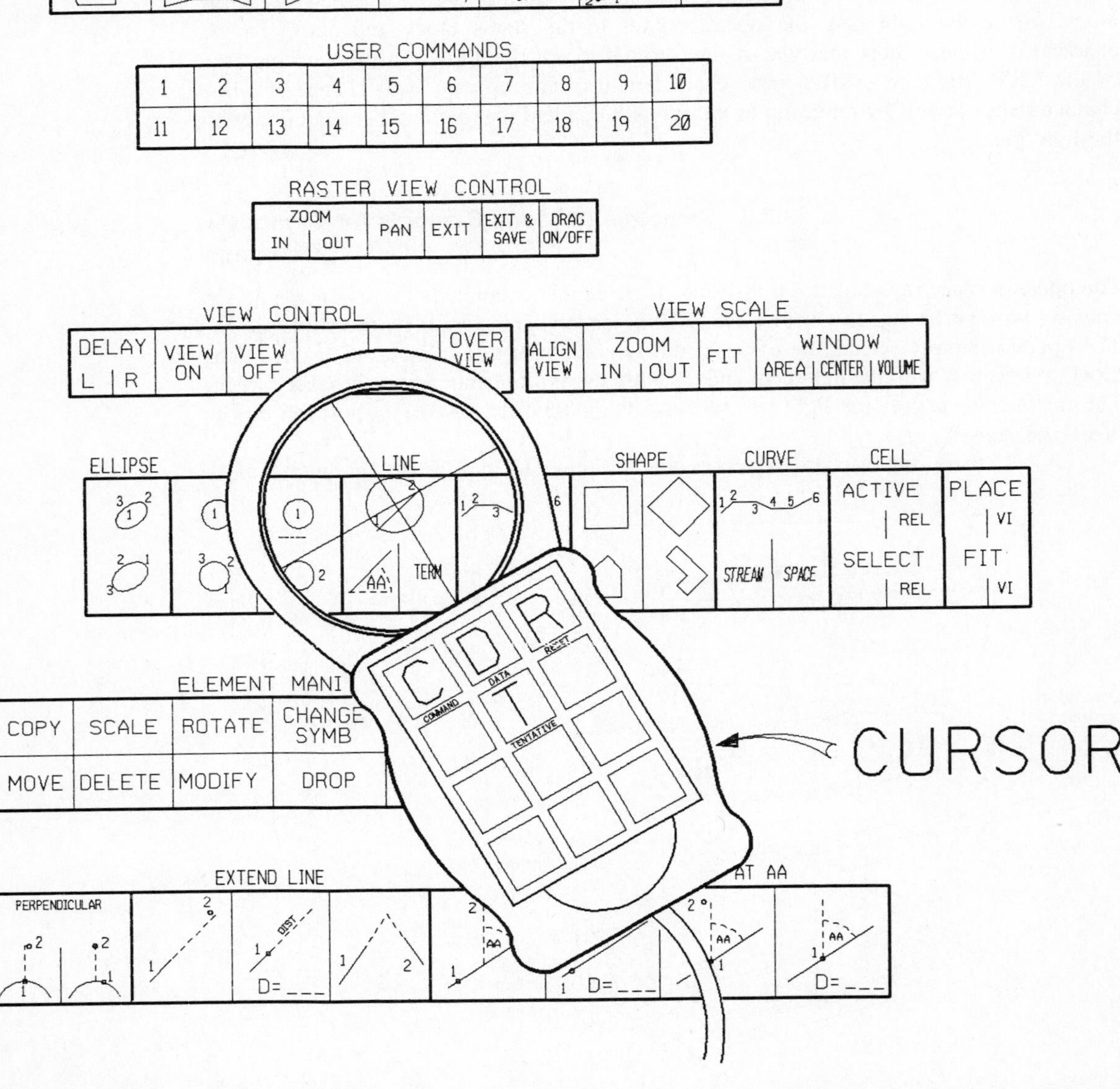


FIGURE 13.3. Typical Menu and Cursor (Courtesy Engineering Graphics Technology, Houston, Texas)

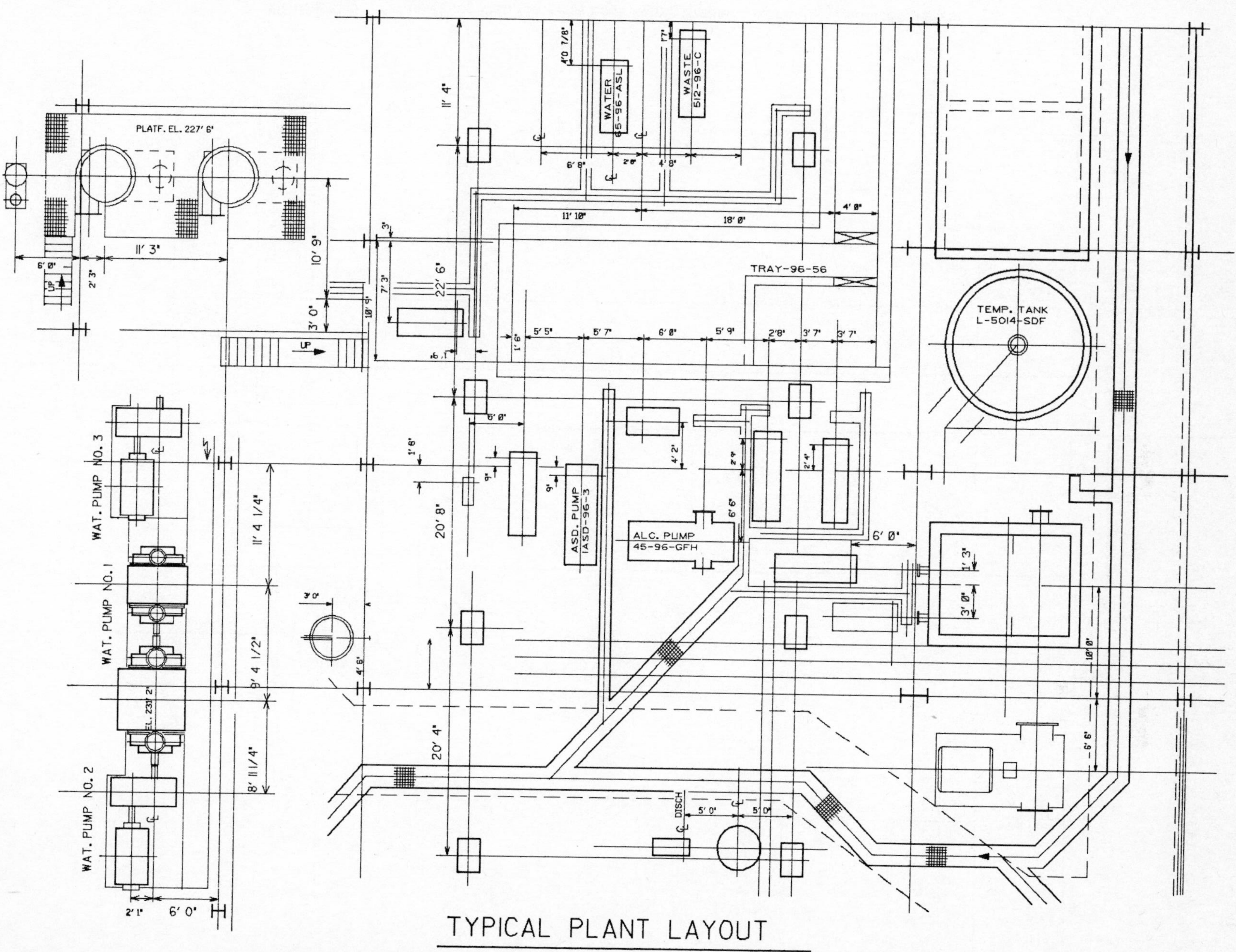


**FIGURE 13.4.** Typical Plant Layout (Courtesy Engineering Graphics Technology, Houston, Texas)

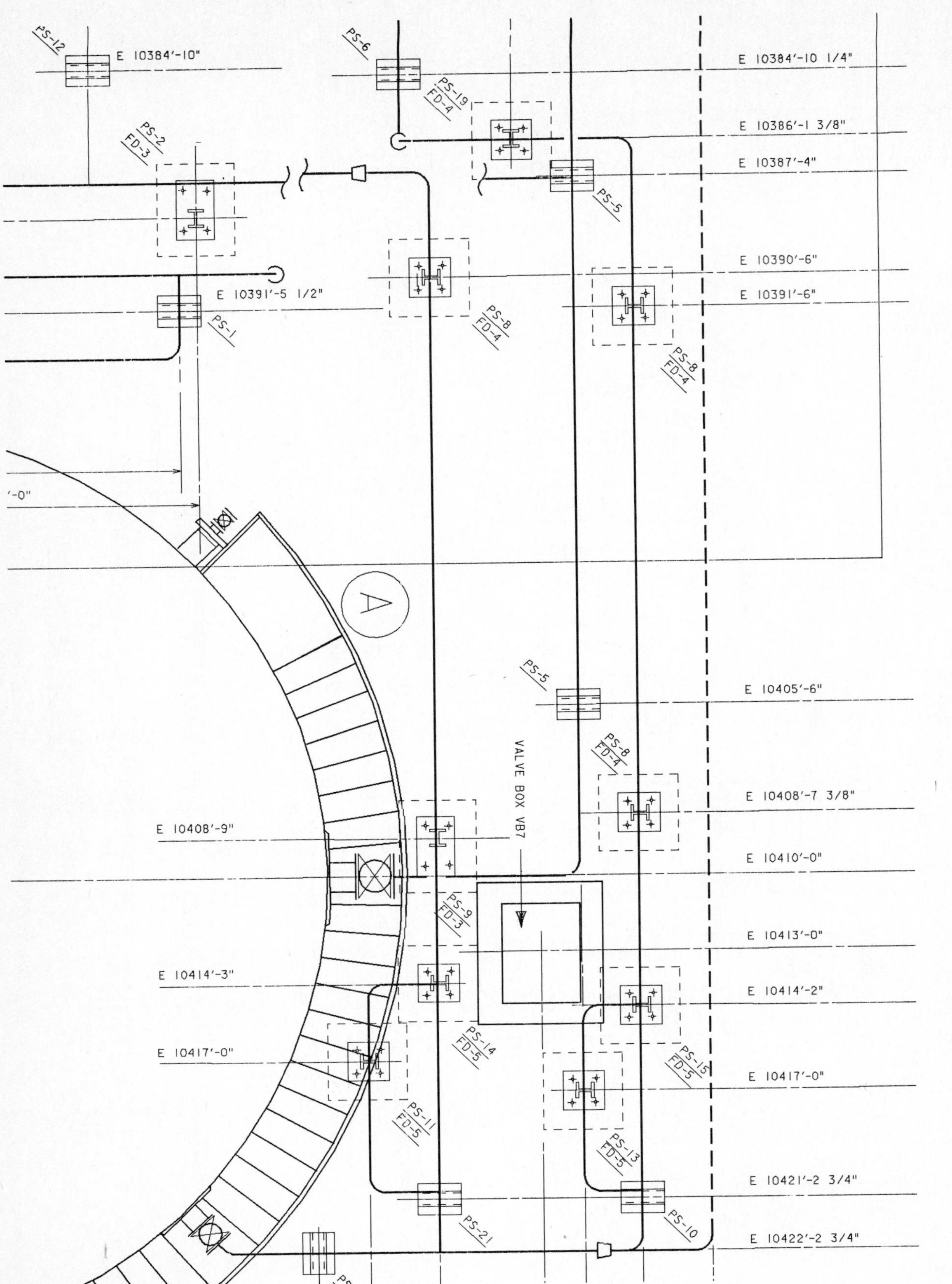


**FIGURE 13.5.** Partial Piping Plan (Courtesy Engineering Graphics Technology, Houston, Texas)

with the light pen and type in the correct data. Changes to drawings recalled from storage can be made in the same way.

With editing capabilities, new numbers, letters, or even sentences can be inserted into the middle of displayed data and the text automatically opens up to make room. Information can be deleted in much the same way with the text closing up to fill in the spaces left by the deleted characters.

### 13.7.9 Layout Changes Accomplished without Redrawing

Should the operator find that portions of the drawing need to be rearranged, there is no need to delete them and start over. Drawings can be changed by moving groups of displayed items, as well as individual symbols, lines, and characters, to any position within the working area of the CRT. The operator simply keys the appropriate switches on the function keyboard and uses the light pen to reposition the data.

### 13.7.10 Displayed Items Copied Automatically

In cases where groups of components or other symbols must be drawn several times to complete a diagram, the operator uses the light pen to point out the items and then types the number of times they are to be redrawn. The system does the rest, automatically drawing the groups of items equidistant from one another and aligning them with the horizontal and vertical axis of the CRT. Figures 13.3, 13.4, and 13.5 show the types of drawings and line work produced by the printer.

### 13.7.11 Adaptable to Engineering Requirements

The highly advanced capabilities of computers can be readily adapted to solve a particular drawing preparation problem. For example, if you have many drawings to produce that are similar in nature, a master drawing can be prepared containing all common symbols, lines, and alphanumerics. The drawing can then be stored on the disc and used as the framework for creating other drawings. In addition, information on each drawing can be supplied by the operator for storage along with the drawing. Typically, the operator enters this information on prepared forms that are called up for viewing on the CRT screen. As an option, these forms can be designed to request only the information specified. In this way, the computer acquires and compiles the exact management program required.

## 13.8 SUMMARY AND CONCLUSION

A recent nationwide survey and study predicts that by 1987 50 to 80 percent of all engineering and architectural drawings will be produced using computer-aided design (CAD) equipment. Take a moment to think about that. We are witnessing a virtual explosion of computer-aided design technology, and it is being applied to down-to-earth everyday production design and drafting to a degree most of us would not have believed possible only a short time ago.

Tremendous opportunities are available today for companies currently utilizing manual design and drafting techniques to attain a quick competitive edge by capitalizing on the productivity gains possible through CAD. Short- and long-term benefits are quite impressive. Productivity ratios from 2–1 to 20–1 and even higher are *very realistically* achievable depending on the drawing type and the skill of the operator. Imagine the impact on a business of achieving even the gains on the conservative end. Such an advantage will be great in the highly competitive engineering market.

# 14 APPENDIX

TABLE OF EXHIBITS

**Exhibit I ABBREVIATIONS: ELECTRICAL SECTION**

| *Abbreviation* | *Unit or Term* | *Abbreviation* | *Unit or Term* |
|---|---|---|---|
| AG | Above grade | BP | Base plate |
| ADJ | Adjustable | BIL | Basic impulse insulation level |
| ABS | Air break switch | BATT | Battery |
| ACB | Air circuit breaker | BCN | Beacon |
| AIR COND | Air conditioning | BRG | Bearing |
| AL | Alarm | BA SW | Bell alarm switch |
| AC | Alternating current | BG | Below grade |
| ALT | Alternator | BM | Bench mark |
| AL | Aluminum | B/M | Bill of material |
| AMB | Amber | BWG | Birmingham Wire Gage |
| AMB | Ambient | BK | Black |
| AMM | Ammeter | BK & W | Black & white |
| AS | Ammeter switch | BLK | Blank |
| AMT | Amount | BOC | Blowout coil |
| AF | Amp fuse | BL | Blue |
| AMP | Ampere | BLR | Boiler |
| AMP HR | Ampere hour | BF | Boiler feed |
| A | AMPS | BFP | Boiler feed pump |
| AT | Ampere trip | BFW | Boiler feed water |
| ANN | Annunciator | BC | Bolt circle |
| ANO | Anode | BOT | Bottom |
| ANT | Antenna | BRKT | Bracket |
| APP | Apparatus | BK | Brake |
| APPX | Appendix | BHP | Brake horsepower |
| APPD | Approved | BMEP | Brake mean effective pressure |
| APPROX | Approximate | BR | Branch |
| ARCH | Architectural | BRZG | Brazing |
| ARM | Armature | BKR | Breaker |
| ARR | Arrester | BRDG | Bridge |
| A.R. | As required | BRT | Brightness |
| ASSY | Assembly | BR STD | British standard |
| ASSN | Association | BTU | British thermal units |
| ATS | Astronomical time switch | BRZ | Bronze |
| ASYM | Asymmetric, asymmetrical | BR | Brown |
| ATM | Atmosphere | B & S | Brown & Sharp (Wire gage same as AWG) |
| ATT | Attach | BLDG | Building |
| ATTEN | Attenuator | BU STD | Bureau of Standards |
| AUD | Audible | BT | Bus tie |
| AUTH | Authorized | BUSH | Bushing |
| AUTO | Automatic | BCT | Bushing current transformer |
| AFC | Automatic frequency control | BUT | Button |
| AUTO RECL | Automatic reclosing | BUZ | Buzzer |
| AUTO XFMR | Autotransformer | BYP | Bypass |
| AUX | Auxiliary | CAB | Cabinet |
| APU | Auxiliary power unit | CA | Cable |
| AVG | Average | CAD | Cadmium |
| AWG | American Wire Gage | CALC | Calculate |
| BC | Back connected | CAL | Calibrate |
| B to B | Back to back | C | Candle |
| BB | Ball bearing | CAND | Candelabra |
| BALL | Ballast | CP | Candlepower |
| BC | Bare copper | CAP | Capacitor, Capacity |
| BAR | Barometer | CARR | Carrier |
| BBL | Barrel | CI | Cast iron |
| BPD | Barrels per day | | |
| BPH | Barrels per hour | | |

**Exhibit I (continued) ABBREVIATIONS: ELECTRICAL SECTION**

| *Abbreviation* | *Unit or Term* | *Abbreviation* | *Unit or Term* |
|---|---|---|---|
| CIP | Cast iron pipe | CNDS | Condensate |
| CAT | Catalog | COND | Condenser |
| C | Celsius, Centigrade | CNDR | Conductor |
| CLG | Ceiling | #/C | Conductor, multiple (Number indicated) |
| CTR | Center | CNDCT | Conductivity |
| CL | Center line | COND OR C | Conduit |
| CG | Center of gravity | CO | Conduit only |
| CM | Centimeter | CONN | Connector, connection |
| CGS | Centimeter-gram-second system | CONST | Constant |
| CENT | Centrifugal | CONT | Contact |
| CH | Chain | CMVM | Contact-making voltmeter |
| CN | Change notice | CNTOR | Contactor |
| CO | Change order | CONTD | Continued |
| CHG (Q) | Charge | CONTR | Contractor |
| CHGR | Charger | CNT | Control, controller |
| CHK | Check | CNT CKT | Control circuit |
| CHEM | Chemical | CPT | Control power transformer |
| CIR | Circle | CSW | Control power switch |
| CKT | Circuit | CR | Control relay |
| CM | Circular mil | CS | Control switch |
| MCM | Circular mils, thousands | CONT | Controller |
| CIRC | Circumference | CONV | Convert |
| CLP | Clamp | COP | Copper |
| CL | Class | CTR | Counter |
| CLR | Clear, clearance | CCW | Counterclockwise |
| CLV | Clevis | CEMF | Counter-electromotive force |
| CW | Clockwise | CPLG | Coupling |
| CCTV | Closed-circuit television | CRK | Crank |
| CL | Closing | XARM | Cross arm |
| CC | Closing coil | XCONN | Cross connection |
| COAX | Coaxial (cable) | XSECT | Cross section |
| COEF | Coefficient | XBAR | Crossbar |
| COLL | Collector | CU | Cubic |
| CFE | Contractor furnished equipment | CC | Cubic centimeter |
| *Color Prefixes* | | CFM | Cubic feet per minute |
| DK | Dark | CFS | Cubic feet per second |
| LT | Light | CU FT | Cubic foot |
| *Colors* | | CU IN | Cubic inch |
| BK | Black | CU M | Cubic meter |
| BL | Blue | CU YD | Cubic yard |
| BR | Brown | CKT | Circuit |
| GRA | Gray | CKT BKR | Circuit breaker |
| GR | Green | CUR | Current |
| OR | Orange | CDR | Current directional relay |
| PU | Purple | CLF | Current limiting fuse |
| RD | Red | CLR | Current limiting resistor |
| SL | Slate (gray) | CT | Current transformer |
| TAN | Tan | CO | Cutout |
| VI | Violet (purple) | CY | Cycle |
| WH | White | | |
| YL | Yellow | | |
| COMB | Combination | | |
| COMM | Communication | | |
| CO | Company | | |
| COMPT | Compartment | | |

**Exhibit I (continued) ABBREVIATIONS: ELECTRICAL SECTION**

| *Abbreviation* | *Unit or Term* |
|---|---|
| CPM | Cycles per minute |
| HZ | Cycles per second |
| CYL | Cylinder |
| DP | Dash pot |
| DB | Decibel |
| DEC | Decimal |
| DK | Deck |
| DEF T | Definite time |
| DTR | Definite time relay |
| DEFL | Deflect |
| (°) DEG | Degree |
| °C | Degrees celsius |
| °F | Degrees fahrenheit |
| DI | Demand indicator |
| DM | Demand meter |
| DEPT | Department |
| DSGN | Design |
| DET | Detail |
| DEV | Develop |
| DP | Dew point |
| DIAG | Diagram |
| DIA | Diameter |
| DIAPH | Diaphragm |
| DENG | Diesel engine |
| D.O. | Diesel oil |
| DIFF | Differential |
| DIFF TR | Differential time relay |
| DIM | Dimension |
| DIO | Diode |
| DIR CONN | Direct connected |
| DC | Direct current |
| DF | Direction finder |
| DIR | Directional |
| DISCH | Discharge |
| DISC | Disconnect |
| DISC SW | Disconnect switch |
| DIST | Distance |
| DIST PNL | Distribution panel |
| DIV | Division |
| DSW | Door switch |
| DBL | Double |
| DG | Double groove (insulators) |
| DGDP | Double groove, double petticoat (insulators) |
| DP SW | Double pole switch |
| DT | Double throw |
| DBLR | Doubler |
| DVTL | Dovetail |
| DWL | Dowel |
| DN | Down |
| DS | Downspout |
| DOZ | Dozen |
| DFTG | Drafting |
| DFSTMN | Draftsman |
| DR | Drain |
| DO | Draw out |
| DWG | Drawing |
| DF | Drinking fountain |
| DP | Drip proof |
| DB | Dry bulb |
| DX | Duplex |
| DT | Dust tight |
| DYNM | Dynamotor |
| EA | Each |
| EB | Ebony asbestos |
| ECC | Eccentric |
| ECON | Economizer |
| EFF | Efficiency |
| ELL | Elbow |
| ELEC | Electric |
| EMH | Electric manhole |
| EWC | Electric water cooler |
| ELECT | Electrical |
| EDP | Electrical distribution panel |
| EMT | Electrical metallic tubing |
| ELECTD | Electrode |
| EM | Electromagnetic |
| ES | Electrostatic |
| ELEM | Elementary |
| EL | Elevation |
| EMER | Emergency |
| EPWR | Emergency power |
| ENCL | Enclosure |
| ENERG | Energized |
| ENG | Engine |
| ENGR | Engineer |
| ENGRG | Engineering |
| ENT | Entrance |
| ENV | Environment, environmental |
| EQ | Equal |
| EQUIP | Equipment |
| EQUIV | Equivalent |
| ESC | Escutcheon |
| EST | Estimate |
| EVAP | Evaporator |
| EXC | Excavate |
| EXCH | Exchange |
| EXC | Excitation |
| EXH | Exhaust |
| EXIST | Existing |
| EXP PR | Explosion proof |
| EXPSD | Exposed |
| EXP | Explosion |
| EXT | External |
| X HVY | Extra heavy |
| X STR | Extra strong |
| EXTR | Extrude |
| FAB | Fabricate |
| F | Fahrenheit |
| FDR | Feeder |
| (')FT | Feet |
| FBM | Feet board measure |
| FPM | Feet per minute |

**Exhibit I (continued) ABBREVIATIONS: ELECTRICAL SECTION**

| *Abbreviation* | *Unit or Term* | *Abbreviation* | *Unit or Term* |
|---|---|---|---|
| FPS | Feet per second | GPM | Gallons per minute |
| FBR | Fiber | GPS | Gallons per second |
| FLD | Field | GALV | Galvanize |
| FIG | Figure | GI | Galvanized iron |
| FIL | Filament | GS | Galvanized steel |
| FIN | Finish | GSWR | Galvanized steel wire rope |
| FA | Fire alarm | GSKT | Gasket |
| FABL | Fire alarm bell | GEN | Generator |
| FABX | Fire alarm box | GFLD | Generator field |
| FCS | Fire control system | GL | Glass |
| FHC | Fire hose cabinet | GOVT | Government |
| FPRF | Fireproof | GFE | Government-furnished equipment |
| FTG | Fitting | GFM | Government-furnished material |
| FIXT | Fixture | GOV | Governor |
| FLG | Flange | GR | Grade |
| FLEX | Flexible | GRAPH | Graphic |
| FLD LT | Flood light | GRTG | Grating |
| FL | Floor | GVL | Gravel |
| FL SW | Flow switch | GR | Green |
| FLUOR | Fluorescent | GROM | Grommet |
| FL MT | Flush mount | GRV | Groove |
| (′)FT | Foot | GND | Ground |
| FC | Foot candle | GR | Group |
| FT LB | Foot pounds | GD | Guard |
| FTG | Footing | GYRO | Gyroscope |
| FD | Forced draft | H-O-A | Hand-off auto |
| FWD | Forward | HR | Hand reset |
| FDN | Foundation | HH | Handhole |
| FRAC | Fractional | HDL | Handle |
| FHP | Fractional horsepower | HGR | Hanger |
| FR | Frame | HD | Hard-drawn |
| FRT | Freight | HDW | Hardware |
| FREQ | Frequency | HAZ | Hazardous |
| HF | Frequency, high | HD | Head |
| LF | Frequency, low | HT | Heat |
| MF | Frequency, medium | HTR | Heater |
| FM | Frequency modulation | H & V | Heating & ventilating |
| SHF | Frequency, super high | HVY | Heavy |
| UHF | Frequency, ultrahigh | HT | Height |
| VHF | Frequency, very high | HZ | Hertz |
| VLF | Frequency, very low | HPFS | High point finished surface |
| FR | Front | HIPOT | High potential test |
| FLA | Full-load current (amps) | HP | High pressure |
| FV | Full voltage | HS | High speed |
| FVNR | Full voltage nonresistant | HSS | High-speed steel |
| FURN | Furnish | HT | High tension |
| FU | Fuse | HV | High voltage |
| FB | Fuse block | HWY | Highway |
| FU BX | Fuse box | HC | Holding coil |
| FSBL | Fusible | HOR | Horizontal |
| FUT | Future | HP | Horsepower |
| GA | Gage or gauge | HOSP | Hospital |
| GAL | Gallon | H | Hot (wire) |
| GPH | Gallons per hour | HR | Hour |

**Exhibit I (continued) ABBREVIATIONS: ELECTRICAL SECTION**

| *Abbreviation* | *Unit or Term* |
|---|---|
| HSG | Housing |
| CWT | Hundredweight |
| HYD | Hydraulic |
| HYDRO | Hydrostatic |
| IDENT | Identify, identification |
| IGN | Ignition |
| ILLUM | Illuminate |
| ILLUS | Illustrate |
| IMP | Impedance |
| INCAND | Incandescent |
| (") IN. | Inch |
| INC | Incoming |
| IS | Incomplete sequence (relay) |
| INDEP | Independent |
| I/L | Indicating light, colored (color indicated) example |
| IND | Indicator |
| IND LP | Indicator lamp |
| IND | Induction |
| INFL | Inflammable |
| INFO | Information |
| ID | Inside diameter |
| INSTL | Installation |
| INST | Instantaneous |
| INST RLY | Instantaneous relay |
| IB | Instruction book |
| INSTR | Instrument |
| INSTM | Instrumentation |
| INS | Insulate |
| IT | Insulating transformer |
| IC | Intercommunication |
| INTCON | Interconnection |
| INT | Interior |
| IC | Interior communication |
| INTLK | Interlock |
| INTMT | Intermittent |
| INTL | International |
| IACS | International Annealed copper Standard |
| IPS | International pipe standard |
| SI | International system of units |
| ICS | Interphone control station |
| INTER | Interrupt |
| IC | Interrupting capacity |
| INV | Inverse |
| INTR | Inverse time relay inverter |
| JKT | Jacket |
| JW | Jacket Water |

| *Abbreviation* | *Unit or Term* |
|---|---|
| JT | Joint |
| JAN | Joint Army–Navy |
| JCT | Junction |
| JB | Junction Box |
| K | Kelvin |
| KA | Kiloampere |
| KC | Kilocycle |
| KC | Kilocycles per second |
| KG | Kilogram |
| KHZ (kH$_z$) | Kilohertz |
| KM | Kilometer |
| KVAR | Kilovar |
| KV | Kilovolt |
| KVA | Kilovolt-ampere |
| KVAH | Kilovolt-ampere hour |
| KVAR | Kilovolt-ampere reactive (capacitive) |
| KW | Kilowatt |
| KWH | Kilowatt hour |
| KO | Knockout |
| LAB | Laboratory |
| L | Lamp |
| GIL | Lamp, green indicating |
| RIL | Lamp, red indicating |
| YIL | Lamp, yellow indicating |
| WIL | Lamp, white indicating |
| LC | Load center |
| LDG | Loading |
| LAT | Latitude |
| LAV | Lavatory |
| LC | Lead covered |
| LG | Length |
| LTR | Letter |
| LA | Level arm |
| LIC | License |
| LT | Light |
| LTG | Lighting |
| LA | Lightning arrester |
| LIM | Limit |
| LS | Limit switch |
| LIQ | Liquid |
| LF | Load factor (tabular only) |
| LLR | Load-limiting resistor |
| LR | Load ratio |
| LOS | Lock-out stop |
| LRS | Locked rotor current (amps) |
| LKR | Locker |
| LOG | Logarithm |
| LG | Long |
| LONG | Longitude |

**Exhibit I (continued) ABBREVIATIONS: ELECTRICAL SECTION**

| *Abbreviation* | *Unit or Term* |
|---|---|
| LP | Low pressure |
| LT | Low tension |
| LV | Low voltage |
| LVR | Low-voltage release |
| LS | Low speed |
| LT | Low torque |
| LUB | Lubricate |
| L | Lumen |
| LPW | Lumens per watt |
| MACH | Machine |
| MAG | Magnet, magnetic |
| MMF | Magnetomotove force |
| MN | Main |
| MDS | Main distributing frame |
| MAINT | Maintenance |
| MALL | Malleable |
| MI | Malleable iron |
| MH | Manhole |
| MAN | Manual |
| MVC | Manual volume control |
| MFR | Manufacturer |
| MFD | Manufactured |
| MFG | Manufacturing |
| MK | Mark |
| MA | Master |
| MSW | Master switch |
| MATL | Material |
| MTO | Material take off |
| MAX | Maximum |
| MEP | Mean effective pressure |
| MECH | Mechanical |
| MED | Medium |
| MC | Megacycles |
| MV | Megavolt |
| MVA | Megavolt ampere |
| MW | Megawatt |
| MWH | Megawatt-hour |
| MEG | Megohm |
| MEMO | Memorandum |
| MV | Mercury vapor |
| MET | Metal |
| MEZZ | Mezzanine |
| μA or UA | Microampere |
| μF or UF | Microfarad |
| μH or UH | Microhenry |
| μs | Microsiemens |
| MIKE | Microphone |
| μs or US or S | Microsecond |
| μV or UV | Microvolt |
| μV/M or UV/M | Microvolts per meter |
| μW or UW | Microwatt |
| MI | Miles |
| MPG | Miles per gallon |
| MPH | Miles per hour |
| MPHPS | Miles per hour per second |
| MIL-STD | Military Standard (book) |
| MA | Milliampere |
| MG | Milligram |
| MH | Millihenry |
| ML | Millilambert |
| MM | Millimeter |
| MCM | Million circular mills |
| MGD | Million gallons per day |
| MS | Milliseconds |
| mV | Millivolt |
| mW | Milliwatt |
| MIN | Minimum |
| (′) MIN | Minute |
| MISC | Miscellaneous |
| MIX | Mixture |
| MOD | Modify |
| MOL WT | Molecular weight |
| MOM | Momentary |
| MC | Momentary contact |
| MO | Month |
| MON | Monument |
| MTR | Motor |
| MCP | Motor circuit protector |
| MCC | Motor control center |
| MF | Motor field |
| MG | Motor generator |
| MTD | Mounted |
| MTG | Mounting |
| MULT | Multiple |
| NP | Nameplate |
| NEC | National Electric Code |
| NESC | National Electric Safety Code |
| NEMA | National Electric Manufacturing Association |
| NEG | Negative |
| NET | Network |
| NEU | Neutral |
| NVR | No voltage release |
| NM | Nonmetallic |
| NMAG | Nonmagnetic |
| NOM | Nominal |
| NOR | Normal |
| NC | Normally closed |
| NO | Normally open |
| N | North |
| NA | Not applicable |
| NF | Not fused |
| NIC | Not in contact |
| NTS | Not to scale |
| NO | Number |

**Exhibit I (continued) ABBREVIATIONS: ELECTRICAL SECTION**

| *Abbreviation* | *Unit or Term* |
|---|---|
| OBS | Obsolete |
| OCT | Octagon |
| OFF | Office |
| Ω | Ohm |
| OCB | Oil circuit breaker |
| OI | Oil insulated |
| OISC | Oil insulated, self-cooled |
| OIFC | Oil insulated, fan-cooled |
| OS | Oil switch |
| OT | Oil tight |
| ODP | Open drip proof |
| OPNG | Opening |
| OPR | Operate |
| OPP | Opposite |
| OPT | Optical |
| O | Orange |
| ORIG | Original |
| OSC | Oscillate |
| OZ | Ounce |
| OUT | Outlet |
| OD | Outside diameter |
| OC | Overcurrent |
| OVHD | Overhead |
| O.L. | Overload |
| OVV | Overvoltage |
| PTD | Painted |
| PR | Pair |
| PNL | Panel |
| PAR | Parallel |
| PKWY | Parkway |
| PT | Part |
| PPM | Parts per million |
| PTD | Partition |
| PTY | Party |
| PASS | Passenger |
| PAT | Patent |
| PRD | Pedestal |
| PEND | Pendent |
| PCT | Percent |
| % IMPD | Percent impedence |
| PERF | Perforate |
| PERM | Permanent |
| PM | Permanent magnet |
| PERP | Perpendicular |
| PH φ | Phase |
| PHEN | Phenolic |
| PU | Pickup |
| pF | Picofarad |
| PIX | Picture |
| PC MK | Piece mark |
| PLT | Pilot |
| PT | Pint |
| P & ID | Piping & instrument diagram |
| PLSTC | Plastic |
| PLATF | Platform |
| PL | Plated |
| PNEU | Pneumatic |
| PT | Point |
| POL | Polarized |
| PVC | Polyvinyl chloride |
| PORC | Porcelain |
| PORT | Portable |
| POS | Positive |
| POT | Potentiometer, potential |
| PT | Potential transformer |
| LB | Pound |
| PCF | Pounds per cubic foot |
| PSF | Pounds per square foot |
| PSI | Pounds per square inch |
| PSIA | Pounds per square inch absolute |
| PWR | Power |
| PA | Power amplifier |
| PDR | Power directional relay |
| PF | Power factor |
| PFM | Power factor meter |
| PP | Power plant |
| PWR SPLY | Power supply |
| PRCST | Precast |
| PREFAB | Prefabricated |
| PFD | Preferred |
| PREP | Prepare |
| PRESS | Pressure |
| | Pressure switch |
| PRI | Primary |
| PAX | Private automatic exchange |
| PBX | Private branch exchange |
| PROC | Process |
| PROD | Production |
| PROT | Protection |
| PA | Public address |
| PUB | Publication |
| PB | Pull box |
| PULL B SW | Pull-button switch |
| PUR | Purchase |
| PB | Push button |
| PB STA | Push-button station |
| P-P | Push–pull |
| QUAD | Quadrant |
| QUAL | Quality |
| QTY | Quantity |
| QT | Quart |
| QTR | Quarter |
| QTZ | Quartz |
| RAD | Radial |
| RDF | Radio direction finding |
| RF | Radio frequency |

**Exhibit I (continued) ABBREVIATIONS: ELECTRICAL SECTION**

| *Abbreviation* | *Unit or Term* |
|---|---|
| RFI | Radio-frequency interference |
| R | Radius |
| RR | Railroad |
| | Raintight |
| REAC | Reactive, reactant, reactor |
| RFM | Reactive factor meter |
| KVAR | Reactive kilovolt ampere |
| VAR | Reactive volt ampere |
| RVM | Reactive voltmeter |
| REAC | Reactor |
| RECD | Received |
| RECP | Receptacle |
| RECL | Reclosing |
| RDM | Recording demand meter |
| RECT | Rectifier |
| R | Red |
| RED | Reducer |
| REF | Reference |
| REFL | Reflector |
| REGEN | Regenerative |
| REG | Regulator |
| REINF | Reinforce |
| REBAR | Reinforcing steel |
| RH | Relative humidity |
| REL | Relay |
| RC | Remote control |
| REM | Remove |
| REN | Renewable |
| REP | Repair |
| REPL | Replace |
| REPRO | Reproduce |
| REQD | Required |
| REQ | Requisition |
| RESID | Residual |
| RES | Resistance |
| RTD | Resistance temperature detector |
| RES | Resistor |
| REV | Reverse |
| REV | Revolution |
| RPM | Revolutions per minute |
| RPS | Revolutions per second |
| RHEO | Rheostat |
| R | Right |
| R/W | Right of way |
| R | Right |
| RGS | Rigid galvanized steel |
| RING | Ringing |
| RD | Road |
| RB | Roller bearing |
| RF | Roof |
| RM | Room |
| RD | Root diameter |
| RMS | Root mean square |
| ROT | Rotary |
| RD | Round |
| SAF | Safety |
| SD BL | Sand blast |
| SAT | Saturate |
| SSF | Saybolt seconds furol (oil viscosity) |
| SSU | Saybolt seconds universal (oil viscosity) |
| SCHED | Schedule |
| SCHEM | Schematic |
| SMLS | Seamless |
| SL | Searchlight |
| SEC | Second |
| SEC | Secondary |
| SECT | Section |
| SEL | Selector |
| SEP | Separate |
| SEQ | Sequence |
| | Series |
| SFT | Shaft |
| SHT | Sheet |
| SHLD | Shield |
| | Short circuit |
| SW | Short wave |
| SH BL | Shot blast |
| SH | Shunt |
| SIG | Signal |
| SIL | Silence |
| SCR | Silicon-controlled rectifier |
| SIM | Similar |
| SX | Simplex |
| S | Single |
| SF | Single feeder |
| SG | Single groove (insulators) |
| SGSP | Single groove, single petticoat (insulators) |
| SPDT | Single pole, double throw |
| SPST | Single pole, single throw |
| ST | Single throw |
| SK | Sketch |
| SLV | Sleeve |
| SB | Sleeve bearing |
| SML | Small |
| SMK | Smoke |
| SKT | Socket |
| SLD | Solder |
| SOL | Solenoid |
| SOV | Solenoid valve |
| SN | Solid neutral |
| SND | Sound |

**Exhibit I (continued) ABBREVIATIONS: ELECTRICAL SECTION**

| *Abbreviation* | *Unit or Term* |
|---|---|
| SPH | Space heater |
| SP | Spare |
| SPKR | Speaker |
| SPL | Special |
| SP | Specific |
| SP GR | Specific gravity |
| SPEC | Specification |
| SP | Speed |
| SP | Splash proof |
| SP PH | Split phase |
| SQ | Square |
| SQ CG | Squirrel cage |
| STN | Stainless |
| SST | Stainless steel |
| STAN | Stanchion |
| STD | Standard |
| ST | Start |
| SS | Start–stop |
| STR | Starter |
| STG | Starting |
| STA | Station |
| STL | Steel |
| SP | Stop |
| STG | Storage |
| STR | Straight |
| STRD | Stranded |
| ST | Street |
| STRUCT | Structural |
| SB | Stuffing box |
| SUBSTA | Substation |
| SUB | Substitute |
| SUM. | Summary |
| SUPT | Superintendent |
| SUPSD | Supersede |
| SUPP | Supplement |
| SW | Switch |
| *Switch and relay types* | |
| SP SW | Single-pole switch |
| SPST SW | Single-pole, single-throw switch |
| SPDT SW | Single-pole, double throw switch |
| DP SW | Double-pole switch |
| DPST SW | Double-pole, single throw switch |
| DPDT SW | Double-pole, double-throw switch |
| 3P SW | Triple-pole switch |
| 3PST SW | Triple-pole, single-throw switch |
| 3PDT SW | Triple-pole, double-throw switch |
| 4P SW | Four-pole switch |
| 4PST SW | Four-pole, single-throw switch |
| 4PDT SW | Four-pole, double-throw switch etc. |
| SWBD | Switchboard |
| SWGR | Switchgear |
| SYM | Symmetrical |
| SYNC | Synchronize |
| SYN | Synchronous |
| SYNSCP | Synchroscope |
| SYS | System |
| TACH | Tachometer |
| TDM | Tandem |
| TAN | Tangent |
| TPR | Taper |
| TECH | Technical |
| TB | Technical bulletin |
| TM | Technical manual |
| TR | Technical report |
| TLG | Telegraph |
| TLM | Telemeter |
| TEL | Telephone |
| TWX | Teletypewriter exchange |
| TV | Television |
| TEMP | Temperature |
| TA | Temperature alarm |
| TC | Temperature compensating |
| TEMP | Template |
| TENS | Tension |
| TERM | Terminal |
| TBD | Terminal board |
| TB | Terminal block |
| TB | Terminal box |
| TER | Tertiary |
| TSW | Test switch |
| IE | That is |
| THEO | Theoretical |
| THRM | Thermal |
| TE | Thermal element |
| TC | Thermocouple |
| THERM | Thermometer |
| THERMO | Thermostat |
| THK | Thick |
| M | Thousand |
| MCM | Thousand circular mills |
| MCF | Thousand cubic feet |
| KIP-FT | Thousand foot pound |
| KIP | Thousand pound |
| THD | Thread |
| TMBR | Timber |
| T | Time |
| TCOD | Time close on de-energization |
| TCOC | Time close on energization |
| TD | Time delay |
| TDC | Time-delay closing |
| TDO | Time-delay opening |
| TOOD | Time open on de-energization |

**Exhibit I (continued) ABBREVIATIONS: ELECTRICAL SECTION**

| *Abbreviation* | *Unit or Term* |
|---|---|
| TOOE | Time open on energization |
| TS | Time switch |
| TC | Timed closing |
| TO | Timed opening |
| TD | Tinned |
| TGL | Toggle |
| TOL | Tolerance |
| TOT | Total |
| TE | Totally enclosed |
| TEFC | Totally enclosed fan cooled |
| TCR | Tracer |
| TNG | Training |
| XCVR | Transceiver |
| XDCR | Transducer |
| TRANS | Transfer |
| XFMR | Transformer |
| XMTR | Transmitter |
| TC | Trip coil |
| TPP | Tripping |
| TRX | Triplex |
| TUB | Tubing |
| TURB | Turbine |
| TURBO GEN | Turbine generator |
| TYP | Typical |
| ULT | Ultimate |
| UHE | Ultrahigh frequency |
| UND GND | Underground |
| UV | Under voltage |
| UVD | Under voltage device |
| UPS | Uninterruptable power systems |
| USG | United States gage |
| USS | United States standard |
| UNIV | Universal |
| UON | Unless otherwise noted |
| VAC | Vacuum |
| VT | Vacuum tube |
| VA | Valve (context) |
| VAP PRF | Vapor proof |
| VS | Vapor seal |
| VT | Vapor tight |
| VAR | Variable |
| VFO | Variable frequency oscillator |
| VEL | Velocity |
| VENT | Ventilate |
| VS | Versus |
| VERT | Vertical |
| VIB | Vibration |
| VISC | Viscosity |
| V | Volt, voltage |
| ADJ R | Voltage-adjusting rheostat |
| VD | Voltage drop |
| VR | Voltage regulator |
| VR | Voltage relay |
| VA | Voltampere (context) |
| VAR | Volt ampere reactive |
| VM | Voltmeter |
| VMSW | Voltmeter switch |
| VPM | Volts per mil |
| VOL | Volume |
| WHSE | Warehouse |
| WCR | Water cooler |
| WH | Water heater |
| WTRPRF | Waterproof |
| WPFC | Waterproof, fan cooled |
| WPG | Waterproofing |
| WT | Watertight |
| W | Watt |
| WPC | Watts per candle |
| WH | Watthour |
| WHM | Watthour meter |
| #WAY | Way, multiple (number indicated) |
| WM | Wattmeter |
| WTHPRF | Weatherproof |
| WP | Weatherproof (insulation) |
| WK | Week |
| WT | Weight |
| WB | Wet bulb |
| W | White |
| W | Width |
| WD | Wind |
| WDG | Winding |
| W | Wire |

**Exhibit 2 SUMMARY OF BASIC EQUATIONS USING VOLT, OHM, AMPERE, AND WATT USED IN DIRECT-CURRENT CIRCUITS**

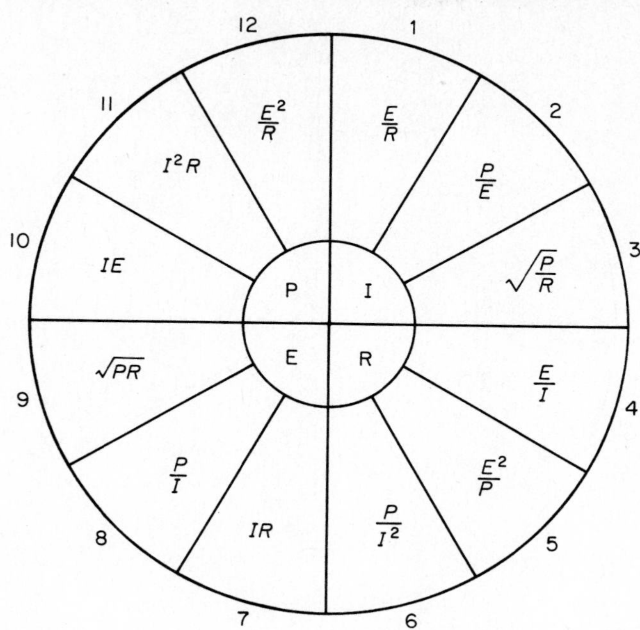


$E$ = voltage
$R$ = resistance in ohms
$I$ = amperes of current
$P$ = power or watts
1000 watts = 1 kilowatt
746 watts = 1 horsepower

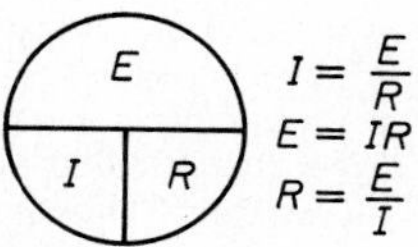


Ohm's law in diagram form

**Exhibit 3 BASIC FORMULAS USING VOLT, OHM, AMPERE, AND WATT USED IN ALTERNATING CURRENT CIRCUITS**

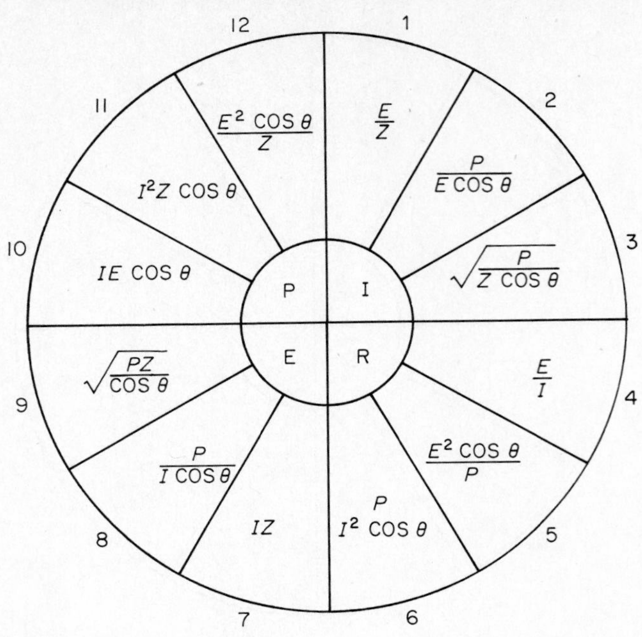


$E$ = voltage
$R$ = resistance in ohms
$I$ = amperes of current
$P$ = power or watts
$\theta$ = phase angle in degrees
1000 watts = 1 kilowatt
746 watts = 1 horsepower

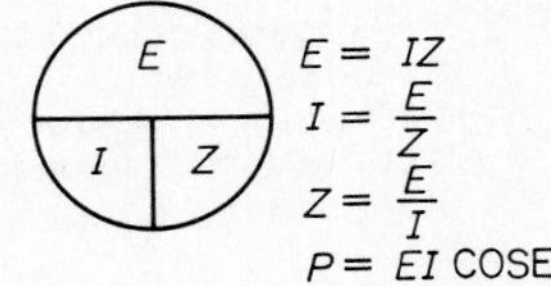


$E = IZ$
$I = \frac{E}{Z}$
$Z = \frac{E}{I}$
$P = EI$ COSE

Ohm's law in diagram form

**Exhibit 4 ELECTRICITY: COMMON FORMULAS**

| *Basic Circuits* | *Units* | *Terms* |
|---|---|---|
| 1. $E = IR$ | Volts = Amps × Ohms | $E$ = Voltage (volts)<br>$I$ = Current (amperes)<br>$R$ = Resistance (ohms)<br>$Z$ = Impedance (ohms)<br>$\pi$ = 3.1416<br>$X_L$ = Industive Reactance (ohms)<br>$f$ = frequency in hertz or cycles per second (cps)<br>$l$ = Inductance (henrys)<br>$C$ = Capacitance (farads)<br>$X_c$ = Capacitive Reactance (ohms) |
| 2. $E = IZ$ | Volts = Amps × Ohms | |
| 3. $X_1 = 2\pi fl$ | Ohms = $2\pi$ × cps × henrys | |
| 4. $X_c = \frac{l}{2\pi fC}$ | Ohms = $\frac{l}{2\pi \times \text{cps} \times \text{farads}}$ | |
| 5. $Z = \sqrt{R^2 + (X_L - X_c)^2}$ | All in ohms | $R_1$ Resistances<br>$R_2$ in<br>$R_3$ Parallel (ohms) |
| 6. $\frac{1}{R_N} = \frac{1}{R_1} + \frac{1}{R_2} + \frac{1}{R_3} \cdots$ | All in ohms | $R_N$ = Equivalent Resistance (ohms) |

| *Power Factor and Power* | *Units* | *Terms* |
|---|---|---|
| 7. PF = $\cos\theta$ | Pure Number | PF = Power Factor (no units)<br>$\theta$ = Power Factor Angle (angle of lead or lead or lag between voltage and current) |
| 8. $\frac{\text{KW}}{\text{KVA}}$ = PF | $\frac{\text{Watts}}{\text{Volts} \times \text{amps}}$ = No Units | KW = Kilowatt<br>KVA = Kilovolt-ampere |
| 9. $P = EI\cos\theta$ (single phase) | Watts = Volts × Amps | P = Power-Watts |
| 10. $P = \sqrt{3}\,EI\cos\theta$ (three phase) | Watts = Volts × Amps | $\sqrt{3}$ = 1.73 |
| 11. $P = I^2R$ (power dissipated) | Watts = Amps × Ohms | |

**Exhibit 5 HANDY FORMULAS FOR DETERMINING KILOWATTS, KVA, REACTIVE KVA, HORSEPOWER, AND AMPERES**

| | *Alternating Current* | | | |
|---|---|---|---|---|
| *Desired Data* | *Single Phase* | *Two Phase*[a] *Four Wire* | *Three Phase* | *Direct Current* |
| Kilowatts | $\frac{\text{Volts} \times \text{amps} \times \text{PF}}{1000}$ | $\frac{2 \times \text{volt} \times \text{amps} \times \text{PF}}{1000}$ | $\frac{1.73 \times \text{volts} \times \text{amps} \times \text{PF}}{1000}$ | |
| kVA | $\frac{\text{Volts} \times \text{amps}}{1000}$ | $\frac{2 \times \text{volts} \times \text{amps}}{1000}$ | $\frac{1.73 \times \text{volts} \times \text{amps}}{1000}$ | $\frac{\text{Volts} \times \text{amps}}{1000}$ |
| Reactive kVA | $\frac{\text{Volts} \times \text{amps} \times \sqrt{1 - \text{PF}^2}}{1000}$ | $\frac{2 \times \text{volts} \times \text{amps} \sqrt{1 - \text{PF}^2}}{1000}$ | $\frac{1.73 \times \text{volts} \times \text{amps} \sqrt{1 - \text{PF}^2}}{1000}$ | |
| Horsepower (output) | $\frac{\text{Volts} \times \text{amps} \times \text{eff} \times \text{PF}}{746 \times 100}$ | $\frac{2 \times \text{volts} \times \text{amps} \times \text{eff} \times \text{PF}}{746 \times 100}$ | $\frac{1.73\ \text{volts} \times \text{amps} \times \text{eff} \times \text{PF}}{746 \times 100}$ | $\frac{\text{Volts} \times \text{amps}}{746 \times 100}$ |
| Amperes (when horsepower is known) | $\frac{\text{HP} \times 746 \times 100}{\text{Volts} \times \text{eff} \times \text{PF}}$ | $\frac{\text{HP} \times 746 \times 100}{2 \times \text{volts} \times \text{eff} \times \text{PF}}$ | $\frac{\text{HP} \times 746 \times 100}{1.73 \times \text{volts} \times \text{eff} \times \text{PF}}$ | $\frac{\text{HP} \times 746 \times 100}{\text{Volts} \times \text{eff}}$ |
| Amperes (when kilowatts are known) | $\frac{\text{Kilowatts} \times 1000}{\text{Volts} \times \text{PF}}$ | $\frac{\text{Kilowatts} \times 1000}{2 \times \text{volts} \times \text{PF}}$ | $\frac{\text{Kilowatts} \times 1000}{1.73 \times \text{volts} \times \text{PF}}$ | $\frac{\text{Kilowatts} \times 1000}{\text{volts}}$ |
| Amperes (when kVA is known) | $\frac{\text{kVA} \times 1000}{\text{Volts}}$ | $\frac{\text{kVA} \times 1000}{2 \times \text{volts}}$ | $\frac{\text{kVA} \times 1000}{1.73 \times \text{volts}}$ | |

[a] *In three-wire, two-phase circuits the current in the common conductor is 1.41 times that in either other conductor.*
*Efficiency (eff) in these formulas is expressed in percent: 95%.*
*Power factor (PF) in these formulas is expressed as a decimal: 0.85.*

## Exhibit 6 EMPTY CONDUIT WEIGHT PER 100-FOOT LENGTH

| *Trade Size* | *Conduit ID* | *Rigid Stl Galv* | | *Rigid Alum* | | *PVC Sch 40* | | *PVC Sch 80* | | *EMT* | | *IMC* | | |
|---|---|---|---|---|---|---|---|---|---|---|---|---|---|---|
| | | *OD* | *WT* | *OD* | *WT* | *OD* | *Wt* | *OD* | *Wt* | *OD* | *Wt* | *ID* | *OD* | *Wt* |
| 1/2 | 0.622 | 0.840 | 79 | 0.840 | 30 | 0.840 | 16 | — | — | 0.706 | 30 | 0.675 | 0.815 | 60 |
| 1/2 | 0.546 | — | — | — | — | — | — | 0.840 | 19 | — | — | — | — | — |
| 3/4 | 0.824 | 1.050 | 105 | 1.050 | 40 | 1.050 | 21 | — | — | 0.922 | 47 | 0.879 | 1.029 | 82 |
| 3/4 | 0.742 | — | — | — | — | — | — | 1.050 | 26 | — | — | — | — | — |
| 1 | 1.049 | 1.315 | 153 | 1.315 | 60 | 1.315 | 31 | — | — | 1.163 | 68 | 1.120 | 1.290 | 116 |
| 1 | 0.957 | — | — | — | — | — | — | 1.315 | 38 | — | — | — | — | — |
| 1 1/4 | 1.380 | 1.660 | 201 | 1.660 | 81 | 1.660 | 42 | — | — | 1.510 | 100 | 1.468 | 1.638 | 150 |
| 1 1/4 | 1.278 | — | — | — | — | — | — | 1.660 | 53 | — | — | — | — | — |
| 1 1/2 | 1.610 | 1.900 | 249 | 1.900 | 97 | 1.900 | 50 | — | — | 1.740 | 114 | 1.703 | 1.883 | 182 |
| 1 1/2 | 1.500 | — | — | — | — | — | — | 1.900 | 64 | — | — | — | — | — |
| 2 | 2.067 | 2.375 | 334 | 2.375 | 132 | 2.375 | 67 | — | — | 2.197 | 147 | 2.170 | 2.360 | 242 |
| 2 | 1.939 | — | — | — | — | — | — | 2.375 | 88 | — | — | — | — | — |
| 2 1/2 | 2.469 | 2.875 | 527 | 2.875 | 208 | 2.875 | 107 | — | — | — | — | 2.597 | 2.857 | 401 |
| 2 1/2 | 2.323 | — | — | — | — | — | — | 2.875 | 134 | — | — | — | — | — |
| 2 1/2 | 2.731 | — | — | — | — | — | — | — | — | 2.875 | 230 | — | — | — |
| 3 | 3.068 | 3.500 | 690 | 3.500 | 272 | 3.500 | 140 | — | — | — | — | 3.216 | 3.476 | 493 |
| 3 | 2.900 | — | — | — | — | — | — | 3.500 | 175 | — | — | — | — | — |
| 3 | 3.356 | — | — | — | — | — | — | — | — | 3.500 | 270 | — | — | — |
| 3 1/2 | 3.548 | 4.000 | 831 | 4.000 | 327 | 4.000 | 169 | — | — | — | — | 3.711 | 3.971 | 573 |
| 4 | 4.026 | 4.500 | 982 | 4.500 | 389 | 4.500 | 199 | — | — | — | — | 4.206 | 4.466 | 638 |
| 4 | 3.826 | — | — | — | — | — | — | 4.500 | 256 | — | — | — | — | — |
| 4 | 4.334 | — | — | — | — | — | — | — | — | 4.500 | 400 | — | — | — |
| 5 | 5.047 | 5.563 | 1344 | 5.563 | 529 | — | — | — | — | — | — | — | — | — |
| 6 | 6.065 | 6.625 | 1770 | 6.625 | 787 | — | — | — | — | — | — | — | — | — |

**Exhibit 7 METRIC CONDUIT DIMENSIONS**

| *Conduit ID (mm)* | *Conduit OD (mm)* | *Area ($mm^2$)* | *Conduit ID (in.)* | *Conduit OD (in.)* | *Area ($in.^2$)* |
|---|---|---|---|---|---|
| 13.2 | 15.2 | 136.8 | 0.52 | 0.60 | 0.21 |
| 16.4 | 18.6 | 211.2 | 0.65 | 0.73 | 0.33 |
| 18 | 20.4 | 254.5 | 0.71 | 0.80 | 0.40 |
| 19.9 | 22.5 | 311 | 0.78 | 0.89 | 0.48 |
| 25.5 | 28.3 | 510.7 | 1.00 | 1.11 | 0.79 |
| 34.2 | 37 | 918.6 | 1.35 | 1.46 | 1.43 |
| 44 | 47 | 1520.5 | 1.73 | 1.85 | 2.35 |
| 51 | 54 | 2042.8 | 2.01 | 2.13 | 3.17 |
| 55.8 | 59.3 | 2445.4 | 2.20 | 2.33 | 3.80 |

**EXHIBIT 8 RIGID-STEEL CONDUIT STANDARD AND LARGE RADIUS ELBOW***

| Trade Size, In. | Standard Elbow | | | Elbow Offset for Large Radius Inch | | | | | | | |
|---|---|---|---|---|---|---|---|---|---|---|---|
| | Radius R | Nominal Straight Length | Offset (O) | 12"R 9" (X) | 15"R 9" (X) | 18"R 10" (X) | 24"R 11" (X) | 30"R 11" (X) | 36"R 11" (X) | 42"R 12" (X) | 48"R 12" (X) |
| 1/2 | 4 | 2.688 | 6.688 | | | | | | | | |
| 3/4 | 4 1/2 | 2.906 | 7.406 | | | | | | | | |
| 1 | 5 3/4 | 2.672 | 8.422 | 21 | 24 | 28 | 35 | 41 | 47 | 54 | 60 |
| 1 1/4 | 7 1/4 | 2.625 | 9.875 | 21 | 24 | 28 | 35 | 41 | 47 | 54 | 60 |
| 1 1/2 | 8 1/4 | 3.250 | 11.500 | 21 | 24 | 28 | 35 | 41 | 47 | 54 | 60 |
| 2 | 9 1/2 | 4.281 | 13.781 | 21 | 24 | 28 | 35 | 41 | 47 | 54 | 60 |
| 2 1/2 | 10 1/2 | 4.750 | 15.250 | 21 | 24 | 28 | 35 | 41 | 47 | 54 | 60 |
| 3 | 13 | 7.375 | 20.375 | | 24 | 28 | 35 | 41 | 47 | 54 | 60 |
| 3 1/2 | 15 | 7.813 | 22.813 | | | 28 | 35 | 41 | 47 | 54 | 60 |
| 4 | 16 | 7.000 | 23.000 | | | | 35 | 41 | 47 | 54 | 60 |
| 5 | 24 | 10.933 | 34.938 | | | | | 41 | 47 | 54 | 60 |
| 6 | 30 | 11.433 | 41.438 | | | | | 41 | 47 | 54 | 60 |

**Courtesy National Fire Protection Association*

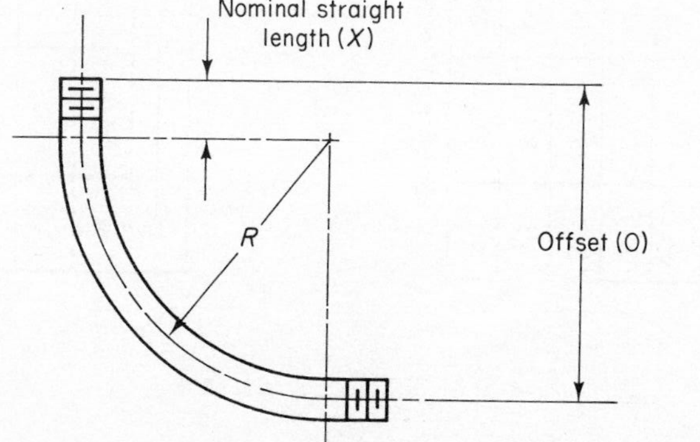

**Exhibit 9 SPACING CHART, MYERS HUBS**

| ↓ | *Conduit Size* → | | | | | | | | | |
|---|---|---|---|---|---|---|---|---|---|---|
| | $\frac{1}{2}$ | $\frac{3}{4}$ | 1 | $1\frac{1}{4}$ | $1\frac{1}{2}$ | 2 | $2\frac{1}{2}$ | 3 | $3\frac{1}{2}$ | 4 |
| $\frac{1}{2}$ | $1\frac{9}{16}$ | — | — | — | — | — | — | — | — | — |
| $\frac{3}{4}$ | $1\frac{11}{16}$ | $1\frac{13}{16}$ | — | — | — | — | — | — | — | — |
| 1 | $1\frac{7}{8}$ | 2 | $2\frac{1}{8}$ | — | — | — | — | — | — | — |
| $1\frac{1}{4}$ | $2\frac{1}{16}$ | $2\frac{3}{16}$ | $2\frac{5}{16}$ | $2\frac{1}{2}$ | — | — | — | — | — | — |
| $1\frac{1}{2}$ | $2\frac{1}{4}$ | $2\frac{3}{8}$ | $2\frac{1}{2}$ | $2\frac{11}{16}$ | $2\frac{7}{8}$ | — | — | — | — | — |
| 2 | $2\frac{1}{2}$ | $2\frac{5}{8}$ | $2\frac{3}{4}$ | $2\frac{15}{16}$ | $3\frac{1}{8}$ | $3\frac{3}{8}$ | — | — | — | — |
| $2\frac{1}{2}$ | $2\frac{3}{4}$ | $2\frac{7}{8}$ | 3 | $3\frac{3}{16}$ | $3\frac{3}{8}$ | $3\frac{5}{8}$ | $3\frac{7}{8}$ | — | — | — |
| 3 | $3\frac{1}{16}$ | $3\frac{3}{16}$ | $3\frac{5}{16}$ | $3\frac{1}{2}$ | $3\frac{11}{16}$ | $3\frac{15}{16}$ | $4\frac{3}{16}$ | $4\frac{1}{2}$ | — | — |
| $3\frac{1}{2}$ | $3\frac{3}{8}$ | $3\frac{1}{2}$ | $3\frac{5}{8}$ | $3\frac{13}{16}$ | 4 | $4\frac{1}{4}$ | $4\frac{1}{2}$ | $4\frac{13}{16}$ | $5\frac{1}{8}$ | — |
| 4 | $3\frac{5}{8}$ | $3\frac{3}{4}$ | $3\frac{7}{8}$ | $4\frac{1}{16}$ | $4\frac{1}{4}$ | $4\frac{1}{2}$ | $4\frac{3}{4}$ | $5\frac{1}{16}$ | $5\frac{3}{8}$ | $5\frac{5}{8}$ |

*Source: Myers Electric Co. Data Sheet.*

**Exhibit 10 SPACING CHART, LOCKNUTS**

| ↓ | *Conduit Size* → | | | | | | | | | |
|---|---|---|---|---|---|---|---|---|---|---|
| | $\frac{1}{2}$ | $\frac{3}{4}$ | 1 | $1\frac{1}{4}$ | $1\frac{1}{2}$ | 2 | $2\frac{1}{2}$ | 3 | $3\frac{1}{2}$ | 4 |
| $\frac{1}{2}$ | $1\frac{1}{4}$ | — | — | — | — | — | — | — | — | — |
| $\frac{3}{4}$ | $1\frac{7}{16}$ | $1\frac{9}{16}$ | — | — | — | — | — | — | — | — |
| 1 | $1\frac{1}{2}$ | $1\frac{11}{16}$ | $1\frac{3}{4}$ | — | — | — | — | — | — | — |
| $1\frac{1}{4}$ | $1\frac{11}{16}$ | $1\frac{7}{8}$ | $1\frac{15}{16}$ | $2\frac{1}{8}$ | — | — | — | — | — | — |
| $1\frac{1}{2}$ | $1\frac{7}{8}$ | 2 | $2\frac{1}{8}$ | $2\frac{15}{16}$ | $2\frac{7}{16}$ | — | — | — | — | — |
| 2 | $2\frac{1}{8}$ | $2\frac{5}{16}$ | $2\frac{3}{8}$ | $2\frac{9}{16}$ | $2\frac{3}{4}$ | 3 | — | — | — | — |
| $2\frac{1}{2}$ | $2\frac{7}{16}$ | $2\frac{5}{8}$ | $2\frac{11}{16}$ | $2\frac{7}{8}$ | $3\frac{1}{16}$ | $3\frac{5}{16}$ | $3\frac{5}{8}$ | — | — | — |
| 3 | $2\frac{7}{8}$ | 3 | $3\frac{1}{8}$ | $3\frac{5}{16}$ | $3\frac{7}{16}$ | $3\frac{3}{4}$ | $4\frac{1}{16}$ | $4\frac{7}{16}$ | — | — |
| $3\frac{1}{2}$ | $3\frac{5}{8}$ | $3\frac{5}{16}$ | $3\frac{3}{8}$ | $3\frac{9}{16}$ | $3\frac{3}{4}$ | 4 | $4\frac{5}{16}$ | $4\frac{3}{4}$ | 5 | — |
| 4 | $3\frac{1}{2}$ | $3\frac{11}{16}$ | $3\frac{3}{4}$ | $3\frac{15}{16}$ | $4\frac{1}{8}$ | $4\frac{3}{8}$ | $4\frac{11}{16}$ | $5\frac{1}{8}$ | $5\frac{3}{8}$ | $5\frac{3}{4}$ |

**Exhibit 11 MINIMUM SPACING CHART FOR C-H EYS SEALS**

| | Conduit Size | | | | | | | | | | | |
|---|---|---|---|---|---|---|---|---|---|---|---|---|
| | 1/2 | 3/4 | 1 | 1 1/4 | 1 1/2 | 2 | 2 1/2 | 3 | 3 1/2 | 4 | 5 | 6 |
| 1/2 | 2 3/8 | 2 21/32 | 3 1/8 | 2 27/32 | 2 31/32 | 3 1/4 | 3 1/2 | 4 1/16 | 4 3/16 | 4 3/8 | 5 1/4 | 6 3/32 |
| 3/4 | 2 21/32 | 2 25/32 | 3 1/4 | 3 1/8 | 3 1/4 | 3 17/32 | 3 25/32 | 4 3/16 | 4 13/32 | 4 21/32 | 5 15/32 | 6 7/32 |
| 1 | 3 1/4 | 3 1/4 | 3 3/8 | 3 19/32 | 3 23/32 | 3 7/8 | 4 1/4 | 4 5/8 | 4 7/16 | 5 1/8 | 5 3/4 | 6 11/32 |
| 1 1/4 | 2 27/32 | 3 1/4 | 3 19/32 | 3 1/16 | 3 9/16 | 3 17/32 | 3 29/32 | 4 17/32 | 5 5/16 | 4 23/32 | 5 13/16 | 6 9/16 |
| 1 1/2 | 2 31/32 | 3 1/4 | 3 23/32 | 3 9/32 | 3 13/32 | 3 31/32 | 4 1/32 | 4 31/32 | 5 3/32 | 5 5/32 | 6 1/4 | 7 |
| 2 | 3 1/4 | 3 17/32 | 3 7/8 | 3 17/32 | 3 31/32 | 3 15/16 | 4 5/16 | 4 15/16 | 5 1/16 | 5 9/16 | 6 7/32 | 6 31/32 |
| 2 1/2 | 3 1/2 | 3 25/32 | 4 1/4 | 3 29/32 | 4 1/32 | 4 5/16 | 4 9/16 | 5 3/16 | 5 5/16 | 5 7/16 | 6 15/16 | 7 7/32 |
| 3 | 4 1/16 | 4 3/16 | 4 5/8 | 4 17/32 | 4 31/32 | 4 15/16 | 5 3/16 | 5 9/16 | 5 13/16 | 6 1/16 | 6 27/32 | 7 19/32 |
| 3 1/2 | 4 3/16 | 4 13/32 | 4 7/16 | 5 5/16 | 5 3/32 | 5 1/16 | 5 5/16 | 5 13/16 | 5 15/16 | 6 | 7 3/32 | 7 27/32 |
| 4 | 4 3/8 | 4 21/32 | 5 1/8 | 4 23/32 | 5 5/32 | 5 9/16 | 5 9/16 | 6 1/16 | 6 | 6 1/4 | 7 11/32 | 8 3/32 |
| 5 | 5 1/4 | 5 15/32 | 5 3/4 | 5 13/16 | 6 1/4 | 6 7/32 | 6 15/16 | 6 27/32 | 7 3/32 | 7 11/32 | 7 31/32 | 8 23/32 |
| 6 | 6 3/32 | 6 7/32 | 6 11/32 | 6 9/16 | 7 | 6 31/32 | 7 7/32 | 7 19/32 | 7 27/32 | 8 3/32 | 8 23/32 | 9 9/32 |

**Exhibit 12 RIGID CONDUIT DIMENSIONS AND PERCENT OF FILL CHART**

| *Conduit Trade Size* | *Conduit ID* | *Area (in.²) 100%* | *Area in Square Inches by Percent of Fill of Conduit* | | | | | | | |
|---|---|---|---|---|---|---|---|---|---|---|
| | | | *60%*[a] | *55%*[b,i] | *53%*[c,i] | *40%*[d] | *38%*[e] | *35%*[f] | *31%*[g] | *30%*[h] |
| 1/2 | 0.622 | 0.30 | 0.18 | 0.17 | 0.16 | 0.12 | 0.12 | 0.11 | 0.09 | 0.09 |
| 3/4 | 0.824 | 0.53 | 0.32 | 0.29 | 0.28 | 0.21 | 0.20 | 0.19 | 0.17 | 0.16 |
| 1 | 1.049 | 0.86 | 0.52 | 0.48 | 0.46 | 0.35 | 0.33 | 0.30 | 0.27 | 0.26 |
| 1 1/4 | 1.380 | 1.50 | 0.90 | 0.82 | 0.79 | 0.60 | 0.57 | 0.52 | 0.46 | 0.45 |
| 1 1/2 | 1.610 | 2.04 | 1.22 | 1.12 | 1.08 | 0.81 | 0.77 | 0.71 | 0.63 | 0.61 |
| 2 | 2.067 | 3.36 | 2.01 | 1.85 | 1.78 | 1.34 | 1.28 | 1.17 | 1.04 | 1.01 |
| 2 1/2 | 2.469 | 4.79 | 2.87 | 2.63 | 2.54 | 1.92 | 1.82 | 1.68 | 1.48 | 1.44 |
| 3 | 3.068 | 7.39 | 4.44 | 4.07 | 3.92 | 2.96 | 2.81 | 2.59 | 2.29 | 2.22 |
| 3 1/2 | 3.548 | 9.89 | 5.93 | 5.44 | 5.24 | 3.95 | 3.76 | 3.46 | 3.06 | 2.97 |
| 4 | 4.026 | 12.73 | 7.64 | 7.00 | 6.75 | 5.09 | 4.84 | 4.46 | 3.95 | 3.82 |
| 4 1/2 | 4.506 | 15.95 | 9.57 | 8.77 | 8.45 | 6.38 | 6.06 | 5.58 | 4.94 | 4.78 |
| 5 | 5.047 | 20.01 | 12.00 | 11.00 | 10.60 | 8.00 | 7.60 | 7.00 | 6.20 | 6.00 |
| 6 | 6.065 | 28.89 | 17.33 | 15.89 | 15.31 | 11.56 | 10.98 | 10.11 | 8.96 | 8.67 |

[a] *60% fill for conduit nipples between enclosures not to exceed 24-in. run length.*

[b] *55% fill for one lead-covered conductor.*

[c] *53% fill for one conductor and conduit stub-up.*

[d] *40% fill for three lead-covered conductors and three or more plain conductors.*

[e] *38% fill for four lead-covered conductors.*

[f] *35% fill for over four lead-covered conductors.*

[g] *31% fill for two plain conductors.*

[h] *30% fill for two lead-covered conductors.*

[i] *Multiconductor cable of three or more conductors shall be treated as a single conductor cable for calculations.*

**Exhibit 13 MULTICONDUCTOR CABLE CONDUIT FILL**

| *7/C #14 Control Cable, Conduit Fill Based on PWC Type 2010, #825-442 OD = 0.510 Area = 0.20 in.²* | | *25 PR #19 Instrument Cable, Conduit Fill Based on PWC Type 150 #621-925 OD = 0.82 Area = 0.573* | | *19/C #14 Control Cable, Conduit Fill Based on PWC Type 2010 #825-522 OD = 0.802 Area = 0.5024* | |
|---|---|---|---|---|---|
| *Cable* | *Conduit (in.)* | *Cable* | *Conduit (in.)* | *Cable* | *Conduit (in.)* |
| 1 | 3/4 | 1 | 1 1/4 | 1 | 1 1/4 |
| 2 | 1 1/4 | 2 | 2 1/2 | 2 | 2 |
| 3 | 1 1/4 | 3 | 3 | 3 | 2 1/2 |
| 4 | 1 1/2 | 4 | 3 | 4 | 3 |
| 5 | 2 | 5 | 3 | 5 | 3 |
| 6 | 2 | 6 | 3 1/2 | 6 | 3 1/2 |
| 7 | 2 1/2 | | | | |
| 8 | 2 1/2 | | | | |
| 9 | 2 1/2 | | | | |
| 10 | 3 | | | | |

**Exhibit 14 DIMENSIONS OF CONDUCTORS**

| *Size AWG MCM* | *Rubber Types RHH, RHW* | | *Simplex CLP Types RHH, RHW* | | *Types THWN, THHN* | | *Types THW, TW* | |
|---|---|---|---|---|---|---|---|---|
| | *OD (in.)* | *Area (in.²)* | *OD (in.)* | *Area (in.²)* | *OD (in.)* | *Area (in.²)* | *OD (in.)* | *Area (in.²)* |
| 14 | 0.204 | 0.0327 | 0.162 | 0.0206 | 0.105 | 0.0087 | 0.162 | 0.0206 |
| 12 | 0.221 | 0.0384 | 0.179 | 0.0251 | 0.122 | 0.0117 | 0.179 | 0.0251 |
| 10 | 0.242 | 0.0460 | 0.199 | 0.0311 | 0.153 | 0.0184 | 0.199 | 0.0311 |
| 8 | 0.311 | 0.0760 | 0.259 | 0.0526 | 0.201 | 0.0317 | 0.259 | 0.0526 |
| 6 | 0.397 | 0.1238 | 0.323 | 0.0819 | 0.257 | 0.0519 | 0.323 | 0.0819 |
| 4 | 0.452 | 0.1605 | 0.372 | 0.1087 | 0.328 | 0.0845 | 0.372 | 0.1087 |
| 3 | 0.481 | 0.1817 | 0.401 | 0.1263 | 0.356 | 0.0995 | 0.401 | 0.1263 |
| 2 | 0.513 | 0.2067 | 0.433 | 0.1473 | 0.388 | 0.1182 | 0.433 | 0.1473 |
| 1 | 0.588 | 0.2715 | 0.508 | 0.2027 | 0.450 | 0.1590 | 0.508 | 0.2027 |
| 1/0 | 0.629 | 0.3107 | 0.549 | 0.2367 | 0.491 | 0.1893 | 0.549 | 0.2367 |
| 2/0 | 0.675 | 0.3578 | 0.595 | 0.2781 | 0.537 | 0.2265 | 0.595 | 0.2781 |
| 3/0 | 0.727 | 0.4151 | 0.647 | 0.3288 | 0.588 | 0.2715 | 0.647 | 0.3288 |
| 4/0 | 0.785 | 0.4840 | 0.705 | 0.3904 | 0.646 | 0.3278 | 0.705 | 0.3904 |
| 250 | 0.868 | 0.5917 | 0.788 | 0.4877 | 0.716 | 0.4026 | 0.788 | 0.4877 |
| 300 | 0.933 | 0.6837 | 0.843 | 0.5581 | 0.771 | 0.4669 | 0.843 | 0.5581 |
| 350 | 0.985 | 0.7620 | 0.895 | 0.6291 | 0.822 | 0.5307 | 0.895 | 0.6291 |
| 400 | 1.032 | 0.8365 | 0.942 | 0.6969 | 0.869 | 0.5931 | 0.942 | 0.6969 |
| 500 | 1.119 | 0.9834 | 1.029 | 0.8316 | 0.955 | 0.7163 | 1.029 | 0.8316 |
| 600 | 1.233 | 1.1940 | 1.143 | 1.0261 | 1.058 | 0.8792 | 1.143 | 1.0261 |
| 700 | 1.304 | 1.3355 | 1.214 | 1.1575 | 1.129 | 1.0011 | 1.214 | 1.1575 |
| 750 | 1.339 | 1.4082 | 1.249 | 1.2252 | 1.163 | 1.0623 | 1.249 | 1.2252 |
| 1000 | 1.494 | 1.7531 | 1.404 | 1.5482 | 1.317 | 1.3623 | 1.404 | 1.5482 |

**Exhibit 15 CIRCUIT CURRENT, SINGLE PHASE**

| *kVA* | *120 V* | *240 V* | *480 V* | *2.4 kV* | *4.16 kV* | *4.8 kV* | *7.2 kV* | *12.47 kV* | *13.2 kV* |
|---|---|---|---|---|---|---|---|---|---|
| 3 | 25 | 12.5 | 6.3 | 1.25 | 0.72 | 0.63 | 0.42 | 0.24 | 0.23 |
| 5 | 41.7 | 20.8 | 10.4 | 2.08 | 1.20 | 1.04 | 0.69 | 0.40 | 0.38 |
| 10 | 83.3 | 41.7 | 20.8 | 4.17 | 2.40 | 2.08 | 1.39 | 0.80 | 0.76 |
| 15 | 125 | 62.5 | 31.3 | 6.25 | 3.61 | 3.13 | 2.09 | 1.21 | 1.14 |
| 25 | 208 | 104 | 52.1 | 10.4 | 6.01 | 5.21 | 3.46 | 2.00 | 1.89 |
| 37.5 | 313 | 157 | 78.1 | 15.6 | 9.01 | 7.81 | 5.2 | 3.01 | 2.84 |
| 50 | 417 | 208 | 104 | 20.8 | 12.0 | 10.4 | 6.9 | 4.01 | 3.79 |
| 75 | 625 | 313 | 156 | 31.3 | 18.0 | 15.6 | 10.4 | 6.02 | 5.68 |
| 100 | 833 | 417 | 208 | 41.7 | 24.0 | 20.8 | 13.9 | 8.03 | 7.58 |
| 167 | 1387 | 694 | 347 | 69.8 | 40.2 | 34.7 | 23.1 | 13.3 | 12.6 |
| 250 | 2083 | 1042 | 521 | 104 | 60.1 | 52.1 | 34.6 | 20.0 | 18.9 |
| 333 | 2775 | 1388 | 694 | 139 | 80 | 69.4 | 46.2 | 26.7 | 25.2 |
| 500 | 4167 | 2083 | 1042 | 208 | 120 | 104 | 69.5 | 40.1 | 37.9 |
| 833 | — | — | 2084 | 417 | 200 | 208 | 115 | 67 | 63 |
| 1250 | — | — | 2605 | 521 | 300 | 261 | 174 | 101 | 95 |
| 1667 | — | — | 3473 | 695 | 401 | 347 | 231 | 133 | 126 |
| 2500 | — | — | 5210 | 1042 | 601 | 521 | 346 | 200 | 189 |
| 3333 | — | — | — | 1389 | 802 | 694 | 462 | 267 | 252 |
| 5000 | — | — | — | 2084 | 1202 | 1042 | 693 | 400 | 378 |

**Exhibit 16 CIRCUIT CURRENT, THREE PHASE**

| *kVA* | *208 V* | *240 V* | *480 V* | *2.4 kV* | *4.16 kV* | *4.8 kV* | *7.2 kV* | *12.47 kV* | *13.2 kV* |
|---|---|---|---|---|---|---|---|---|---|
| 9 | 25.0 | 21.7 | 10.8 | 2.17 | 1.25 | 1.08 | 0.71 | 0.41 | 0.39 |
| 15 | 41.6 | 36.1 | 18.0 | 3.61 | 2.08 | 1.80 | 1.21 | 0.70 | 0.66 |
| 30 | 83.3 | 72.2 | 36.1 | 7.22 | 4.16 | 3.61 | 2.51 | 1.39 | 1.31 |
| 45 | 125 | 108 | 54.1 | 10.8 | 6.25 | 5.41 | 3.61 | 2.09 | 1.97 |
| 75 | 208 | 180 | 90.2 | 18.0 | 10.4 | 9.02 | 6.0 | 3.47 | 3.28 |
| 112.5 | 312 | 271 | 135 | 27.1 | 15.6 | 13.5 | 9.0 | 5.21 | 4.92 |
| 150 | 416 | 361 | 180 | 38.1 | 20.8 | 18.0 | 12.0 | 6.95 | 6.56 |
| 225 | 625 | 541 | 271 | 54.1 | 31.2 | 27.1 | 18.0 | 10.4 | 9.84 |
| 300 | 833 | 722 | 361 | 72.2 | 41.6 | 36.1 | 24.0 | 13.0 | 13.1 |
| 500 | 1388 | 1003 | 601 | 120 | 69.3 | 60.1 | 40.1 | 23.2 | 21.9 |
| 750 | 2082 | 1804 | 902 | 180 | 104 | 90.2 | 60.1 | 34.7 | 32.8 |
| 1000 | 2780 | 2406 | 1203 | 241 | 139 | 120 | 80.7 | 46.6 | 44.0 |
| 1500 | — | 3609 | 1804 | 361 | 208 | 181 | 121 | 70 | 66 |
| 2000 | — | — | 2406 | 480 | 277 | 240 | 159 | 92 | 87 |
| 2500 | — | — | 3007 | 600 | 346 | 300 | 200 | 115 | 109 |
| 3750 | — | — | 4510 | 901 | 520 | 452 | 302 | 175 | 165 |
| 5000 | — | — | — | 1200 | 694 | 600 | 400 | 331 | 218 |

**Exhibit 17 MOTOR FEEDER AND CONTROL REQUIREMENTS BASED ON THW ALUMINUM WIRE AND 3ϕ, 460-VOLT POWER**

| Motor HP | NEMA Starter Size | Motor Full-Load Current (NEC) | Circuit Breaker Setting[a] | Wire Size (THW) | Conduit for 3 Motor Leads | Conduit Size with 3 Motor Leads, and Number of Control Leads Indicated Below | | | | | | | |
|---|---|---|---|---|---|---|---|---|---|---|---|---|---|
| | | | | | | #12 | | | | #14 | | | |
| | | | | | | 2 | 3 | 4 | 5 | 2 | 3 | 4 | 5 |
| 1/2 | 1 | 1.0 | 15 | 12 | 3/4 | 3/4 | 3/4 | 3/4 | 3/4 | 3/4 | 3/4 | 3/4 | 3/4 |
| 3/4 | 1 | 1.4 | 15 | 12 | 3/4 | 3/4 | 3/4 | 3/4 | 3/4 | 3/4 | 3/4 | 3/4 | 3/4 |
| 1 | 1 | 1.8 | 15 | 12 | 3/4 | 3/4 | 3/4 | 3/4 | 3/4 | 3/4 | 3/4 | 3/4 | 3/4 |
| 1 1/2 | 1 | 2.6 | 15 | 12 | 3/4 | 3/4 | 3/4 | 3/4 | 3/4 | 3/4 | 3/4 | 3/4 | 3/4 |
| 2 | 1 | 3.4 | 15 | 12 | 3/4 | 3/4 | 3/4 | 3/4 | 3/4 | 3/4 | 3/4 | 3/4 | 3/4 |
| 3 | 1 | 4.8 | 15 | 12 | 3/4 | 3/4 | 3/4 | 3/4 | 3/4 | 3/4 | 3/4 | 3/4 | 3/4 |
| 5 | 1 | 7.6 | 15 | 12 | 3/4 | 3/4 | 3/4 | 3/4 | 3/4 | 3/4 | 3/4 | 3/4 | 3/4 |
| 7 1/2 | 1 | 11 | 20 | 12 | 3/4 | 3/4 | 3/4 | 3/4 | 3/4 | 3/4 | 3/4 | 3/4 | 3/4 |
| 10 | 1 | 14 | 30 | 12 | 3/4 | 3/4 | 3/4 | 3/4 | 3/4 | 3/4 | 3/4 | 3/4 | 3/4 |
| 15 | 2 | 21 | 50 | 8 | 3/4 | 3/4 | 3/4 | 3/4 | 1 | 3/4 | 3/4 | 1 | 1 |
| 20 | 2 | 27 | 70 | 8 | 3/4 | 3/4 | 3/4 | 3/4 | 1 | 3/4 | 3/4 | 1 | 1 |
| 25 | 2 | 34 | 100 | 6 | 1 | 1 | 1 | 1 | 1 | 1 | 1 | 1 | 1 |
| 30 | 3 | 40 | 100 | 6 | 1 | 1 | 1 | 1 1/4 | 1 1/4 | 1 | 1 | 1 | 1 |
| 40 | 3 | 52 | 150 | 4 | 1 | 1 1/4 | 1 1/4 | 1 1/4 | 1 1/4 | 1 1/4 | 1 1/4 | 1 1/4 | 1 1/4 |
| 50 | 3 | 65 | 175 | 2 | 1 1/4 | 1 1/4 | 1 1/4 | 1 1/4 | 1 1/4 | 1 1/4 | 1 1/4 | 1 1/4 | 1 1/4 |
| 60 | 4 | 77 | 200 | 1 | 1 1/2 | | | | | | | | |
| 75 | 4 | 96 | 250 | 0 | 2 | | | | | | | | |
| 100 | 4 | 124 | 350 | 000 | 2 | Control circuits shall be in individual conduits for all circuits requiring motor leads larger than #2. Minimum size conduit 1 in. if buried. | | | | | | | |
| 125 | 5 | 156 | 400 | 250MCM | 2 1/2 | | | | | | | | |
| 150 | 5 | 180 | 500 | 300MCM | 2 1/2 | | | | | | | | |
| 200 | 5 | 240 | 600 | 350MCM | 3 | | | | | | | | |

[a]Per mfg. for thermal breaker.

**Exhibit 18 MOTOR FEEDER AND CONTROL REQUIREMENTS BASED ON THW COPPER WIRE AND 3ϕ, 460-VOLT POWER**

| Motor HP | NEMA Starter Size | Motor Full-Load Current (NEC) | Circuit Breaker Setting[a] | Wire Size (THW) | Conduit for 3 Motor Leads | Conduit Size with 3 Motor Leads, and Number of Control Leads Indicated Below | | | | | | | |
|---|---|---|---|---|---|---|---|---|---|---|---|---|---|
| | | | | | | #12 | | | | #14 | | | |
| | | | | | | 2 | 3 | 4 | 5 | 2 | 3 | 4 | 5 |
| 1/2 | 1 | 1.0 | 15 | 12 | 3/4 | 3/4 | 3/4 | 3/4 | 3/4 | 3/4 | 3/4 | 3/4 | 3/4 |
| 3/4 | 1 | 1.4 | 15 | 12 | 3/4 | 3/4 | 3/4 | 3/4 | 3/4 | 3/4 | 3/4 | 3/4 | 3/4 |
| 1 | 1 | 1.8 | 15 | 12 | 3/4 | 3/4 | 3/4 | 3/4 | 3/4 | 3/4 | 3/4 | 3/4 | 3/4 |
| 1 1/2 | 1 | 2.6 | 15 | 12 | 3/4 | 3/4 | 3/4 | 3/4 | 3/4 | 3/4 | 3/4 | 3/4 | 3/4 |
| 2 | 1 | 3.4 | 15 | 12 | 3/4 | 3/4 | 3/4 | 3/4 | 3/4 | 3/4 | 3/4 | 3/4 | 3/4 |
| 3 | 1 | 4.8 | 15 | 12 | 3/4 | 3/4 | 3/4 | 3/4 | 3/4 | 3/4 | 3/4 | 3/4 | 3/4 |
| 5 | 1 | 7.6 | 15 | 12 | 3/4 | 3/4 | 3/4 | 3/4 | 3/4 | 3/4 | 3/4 | 3/4 | 3/4 |
| 7 1/2 | 1 | 11 | 20 | 12 | 3/4 | 3/4 | 3/4 | 3/4 | 3/4 | 3/4 | 3/4 | 3/4 | 3/4 |
| 10 | 1 | 14 | 30 | 12 | 3/4 | 3/4 | 3/4 | 3/4 | 3/4 | 3/4 | 3/4 | 3/4 | 3/4 |
| 15 | 2 | 21 | 50 | 10 | 3/4 | 3/4 | 3/4 | 3/4 | 1 | 3/4 | 3/4 | 3/4 | 3/4 |
| 20 | 2 | 27 | 70 | 8 | 3/4 | 3/4 | 3/4 | 3/4 | 1 | 3/4 | 3/4 | 1 | 1 |
| 25 | 2 | 34 | 100 | 8 | 3/4 | 1 | 3/4 | 3/4 | 1 | 3/4 | 3/4 | 1 | 1 |
| 30 | 3 | 40 | 100 | 6 | 1 | 1 | 1 | 1 1/4 | 1 1/4 | 1 | 1 | 1 | 1 |
| 40 | 3 | 52 | 150 | 6 | 1 | 1 | 1 | 1 1/4 | 1 1/4 | 1 | 1 | 1 | 1 1/4 |
| 50 | 3 | 65 | 175 | 4 | 1 | 1 1/4 | 1 1/4 | 1 1/4 | 1 1/4 | 1 1/4 | 1 1/4 | 1 1/4 | 1 1/4 |
| 60 | 4 | 77 | 200 | 2 | 1 1/4 | 1 1/4 | 1 1/4 | 1 1/4 | 1 1/4 | 1 1/4 | 1 1/4 | 1 1/4 | 1 1/4 |
| 75 | 4 | 96 | 250 | 1 | 1 1/4 | | | | | | | | |
| 100 | 4 | 124 | 350 | 00 | 1 1/2 | Control circuits shall be individual | | | | | | | |
| 125 | 5 | 156 | 400 | 000 | 2 | conduits for all circuits requiring | | | | | | | |
| 150 | 5 | 180 | 500 | 0000 | 2 | motor leads larger than #2. | | | | | | | |
| 200 | 5 | 240 | 600 | 350MCM | 2 1/2 | Minimum size conduit 1 in. if buried. | | | | | | | |

[a]Per mfg. for thermal breaker.

**Exhibit 19 MOTOR FEEDER AND CONTROL REQUIREMENTS BASED ON THWN COPPER WIRE AND 3ϕ, 460-VOLT POWER**

| Motor HP | NEMA Starter Size | Motor Full-Load Current (NEC) | Circuit Breaker Setting[a] | Wire Size (THW) | Conduit for 3 Motor Leads | Conduit Size with 3 Motor Leads, and Number of Control Leads Indicated Below | | | | | | | |
|---|---|---|---|---|---|---|---|---|---|---|---|---|---|
| | | | | | | #12 | | | | #14 | | | |
| | | | | | | 2 | 3 | 4 | 5 | 2 | 3 | 4 | 5 |
| 1/2 | 1 | 1.0 | 15 | 12 | 3/4 | 3/4 | 3/4 | 3/4 | 3/4 | 3/4 | 3/4 | 3/4 | 3/4 |
| 3/4 | 1 | 1.4 | 15 | 12 | 3/4 | 3/4 | 3/4 | 3/4 | 3/4 | 3/4 | 3/4 | 3/4 | 3/4 |
| 1 | 1 | 1.8 | 15 | 12 | 3/4 | 3/4 | 3/4 | 3/4 | 3/4 | 3/4 | 3/4 | 3/4 | 3/4 |
| 1 1/2 | 1 | 2.6 | 15 | 12 | 3/4 | 3/4 | 3/4 | 3/4 | 3/4 | 3/4 | 3/4 | 3/4 | 3/4 |
| 2 | 1 | 3.4 | 15 | 12 | 3/4 | 3/4 | 3/4 | 3/4 | 3/4 | 3/4 | 3/4 | 3/4 | 3/4 |
| 3 | 1 | 4.8 | 15 | 12 | 3/4 | 3/4 | 3/4 | 3/4 | 3/4 | 3/4 | 3/4 | 3/4 | 3/4 |
| 5 | 1 | 7.6 | 15 | 12 | 3/4 | 3/4 | 3/4 | 3/4 | 3/4 | 3/4 | 3/4 | 3/4 | 3/4 |
| 7 1/2 | 1 | 11 | 20 | 12 | 3/4 | 3/4 | 3/4 | 3/4 | 3/4 | 3/4 | 3/4 | 3/4 | 3/4 |
| 10 | 1 | 14 | 30 | 12 | 3/4 | 3/4 | 3/4 | 3/4 | 3/4 | 3/4 | 3/4 | 3/4 | 3/4 |
| 15 | 2 | 21 | 50 | 10 | 3/4 | 3/4 | 3/4 | 3/4 | 3/4 | 3/4 | 3/4 | 3/4 | 3/4 |
| 20 | 2 | 27 | 70 | 8 | 3/4 | 3/4 | 3/4 | 3/4 | 3/4 | 3/4 | 3/4 | 3/4 | 3/4 |
| 25 | 2 | 34 | 100 | 8 | 3/4 | 3/4 | 3/4 | 3/4 | 3/4 | 3/4 | 3/4 | 3/4 | 3/4 |
| 30 | 3 | 40 | 100 | 6 | 3/4 | 3/4 | 3/4 | 3/4 | 1 | 3/4 | 3/4 | 3/4 | 3/4 |
| 40 | 3 | 52 | 150 | 6 | 3/4 | 3/4 | 3/4 | 3/4 | 1 | 3/4 | 3/4 | 3/4 | 3/4 |
| 50 | 3 | 65 | 175 | 4 | 1 | 1 | 1 | 1 | 1 | 1 | 1 | 1 | 1 |
| 60 | 4 | 77 | 200 | 2 | 1 | 1 1/4 | 1 1/4 | 1 1/4 | 1 1/4 | 1 1/4 | 1 1/4 | 1 1/4 | 1 1/4 |
| 75 | 4 | 96 | 250 | 1 | 1 1/4 | | | | | | | | |
| 100 | 4 | 124 | 350 | 00 | 1 1/2 | Control circuits shall be individual | | | | | | | |
| 125 | 5 | 156 | 400 | 000 | 1 1/2 | conduits for all circuits requiring | | | | | | | |
| 150 | 5 | 180 | 500 | 0000 | 2 | motor leads larger than #2. | | | | | | | |
| 200 | 5 | 240 | 600 | 350MCM | 2 1/2 | Minimum size conduit 1 in. if buried. | | | | | | | |

[a]*Per mfg. for thermal breaker.*

## Exhibit 20 MOTOR CONDUIT STUB-UP SCHEDULE*

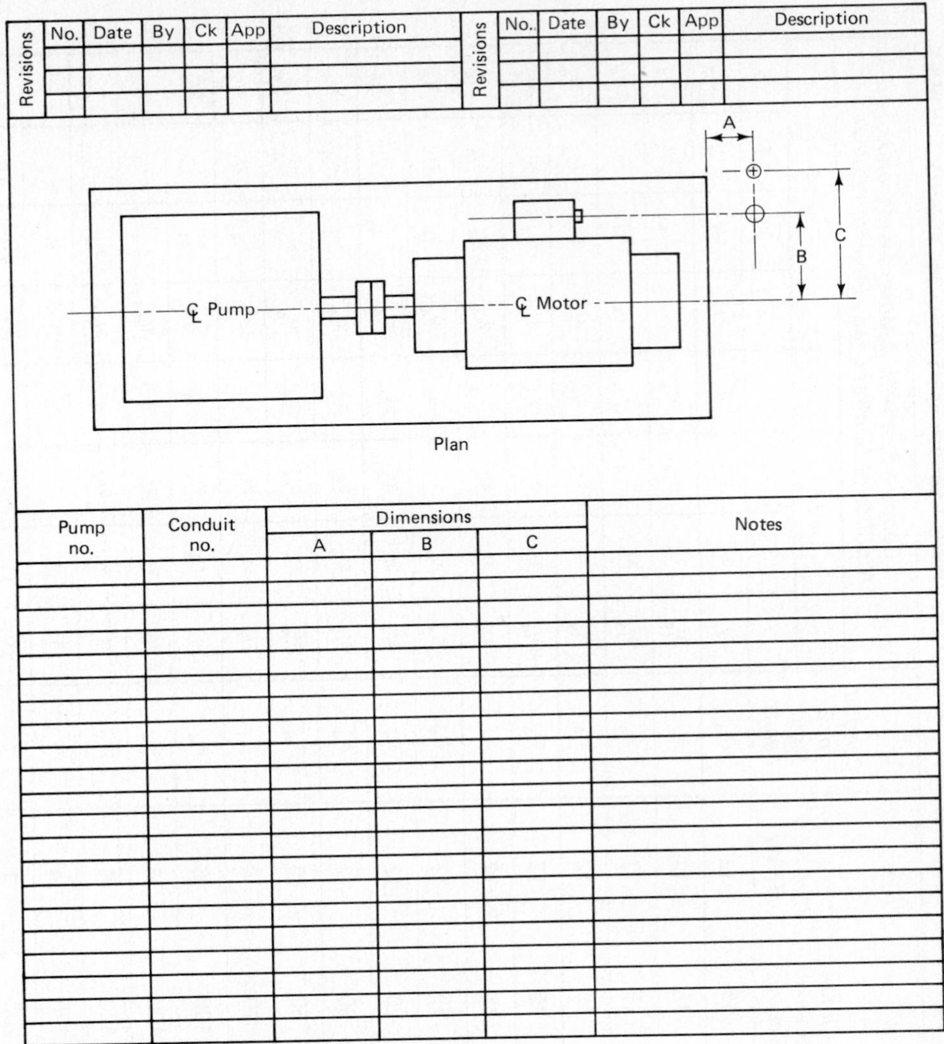

| Revisions | No. | Date | By | Ck | App | Description | Revisions | No. | Date | By | Ck | App | Description |
|---|---|---|---|---|---|---|---|---|---|---|---|---|---|
| | | | | | | | | | | | | | |
| | | | | | | | | | | | | | |
| | | | | | | | | | | | | | |



| Pump no. | Conduit no. | Dimensions | | | Notes |
|---|---|---|---|---|---|
| | | A | B | C | |
| | | | | | |
| | | | | | |
| | | | | | |
| | | | | | |
| | | | | | |
| | | | | | |
| | | | | | |
| | | | | | |
| | | | | | |
| | | | | | |
| | | | | | |
| | | | | | |
| | | | | | |
| | | | | | |
| | | | | | |
| | | | | | |
| | | | | | |
| | | | | | |
| | | | | | |
| | | | | | |

Note: $A = 6\frac{1}{2}$ in. for all conduit sizes up to 4 in. (inclusive)

**Courtesy National Fire Protection Association.*

## Exhibit 21 MOTOR REGISTER*

| Driven equipment | | | Motor driver | | | | | | | | | | | | | | |
|---|---|---|---|---|---|---|---|---|---|---|---|---|---|---|---|---|---|
| Area | Tag number | Service | HP | Syn rpm | Volts | Phase | F.L. amps | Elec type | NEMA design | Encl | Frame size | Type drive | Insul class | Rise C | Motor mfr | P.O. number | marks |

| Revisions | | | | | |
|---|---|---|---|---|---|
| No. | Date | By | Ck | App | Description |

| Revisions | | | | | |
|---|---|---|---|---|---|
| No. | Date | By | Ck | App | Description |

*Courtesy National Fire Protection Association.

**Exhibit 22 NEMA MOTOR ENCLOSURE CHARACTERISTICS**

| *Types* | *Symbol* | *Characteristics* |
|---|---|---|
| *Open* | | |
| Drip proof | ODP | Operates with dripping liquids up to 15° from vertical. |
| Splash proof | OSP | Operates with splashing liquids up to 100° from vertical. |
| Guarded | ODPG | Guarded by limited-size openings (less than $^3/_4$ in.). |
| Semiguarded | DPSG | Only top half of motor guarded. |
| Drip proof fully guarded | DPFG | Drip-proof motor has limited-size openings (less than $^3/_4$ in.). |
| Externally ventilated | OEV | Ventilated with separate motor-driven blower. Can have other types of protection. |
| Pipe ventilated | OPV | Openings accept inlet ducts or pipe for air cooling. |
| Weather protected Type I | WPI | Ventilating passages minimize entrance of rain, snow, and airborne particles; passages less than $^3/_4$ in. in diameter. |
| Weather protected Type II | WPII | In addition to Type I, motors have passages to discharge high-velocity particles blown in. |
| *Totally Enclosed* | | |
| Nonventilated | TENV | Not equipped for external cooling. |
| Fan cooled | TEFC | Cooled by external integral fan. |
| Explosion proof | EP | Withstands internal gas explosion; prevents ignition of external gases. |
| Dust-ignition proof | DUIP | Excludes ignitable amounts of dust and amounts of dust that would degrade performance; prevents ignition of external dusts. |
| Waterproof | TEWP | Excludes leakage except around shaft. |
| Pipe ventilated | TEPV | Openings accept inlet ducts or pipe for air cooling. |
| Water cooled | TEWC | Cooled by circulating water. |
| Water and air cooled | TEWAC | Cooled by water-cooled air. |
| Air to air cooled | TEAC | Cooled by air-cooled air. |
| Guarded TEFC | GTEFC | Fan cooled and guarded by limited-size openings (less than $^3/_4$ in. in diameter). |
| Encapsulated windings | EW | Has resin-filled windings for severe operating conditions; requires one of above enclosures as complete designation (random-wound motor). |
| Sealed windings | SW | Has sealed windings for severe operating conditions; requires one of above enclosures as a complete designation (form-wound motor). |

### Exhibit 23 LIGHTING FIXTURE SCHEDULE

| *Fixture Designation* | *Fixture Description* | *Lamp Type and Wattage* |
|---|---|---|
| | | |
| | | |
| | | |
| | | |
| | | |
| | | |
| | | |
| | | |
| | | |
| | | |
| | | |

### Exhibit 24 NAMEPLATE SCHEDULE*

| Revisions | No. | Date | By | Ck | App | Description | Revisions | No. | Date | By | Ck | App | Description |
|---|---|---|---|---|---|---|---|---|---|---|---|---|---|
| | | | | | | | | | | | | | |
| | | | | | | | | | | | | | |
| | | | | | | | | | | | | | |

| Quantity required | Equipment | Engraving |
|---|---|---|
| | | |
| | | |
| | | |
| | | |
| | | |
| | | |
| | | |
| | | |
| | | |
| | | |
| | | |
| | | |

**Courtesy National Fire Protection Association.*

# INDEX

## C



## D

## E



## F



## G



## H

## I



## J



## K



## L



## M



## N

## O



## P



## R



## S

## T



## U



## V



## W